Joint FAO/WHO Food Standards Program
CODEX ALIMENTARIUS COMMISSION

VOLUME 9A

CODEX ALIMENTARIUS

FISH AND FISHERY PRODUCTS

FOOD AND AGRICULTURE ORGANIZATION OF THE UNITED NATIONS

WORLD HEALTH ORGANIZATION

Rome, 2001

ISBN 92-5-104660-3

THE CODEX ALIMENTARIUS

Preface

The Codex Alimentarius Commission is the international body responsible for the execution of the Joint FAO/WHO Food Standards Programme. Created in 1962 by FAO and WHO the Programme is aimed at protecting the health of consumers and facilitating international trade in foods.

The Codex Alimentarius (Latin, meaning Food Law or Code) is a collection of international food standards adopted by the Commission and presented in a uniform manner. It includes standards for all the principal foods, whether processed or semi-processed or raw. Materials for further processing into foods are included to the extent necessary to achieve the purposes of the Codex Alimentarius as defined. The Codex Alimentarius includes provisions in respect of the hygienic and nutritional quality of food, including microbiological norms, provisions for food additives, pesticide residues, contaminants, labelling and presentations, and methods of analysis and sampling. It also includes provisions of an advisory nature in the form of codes of practice, guidelines and other recommended measures.

This is the second publication of the Codex Alimentarius. The first publication was in 1981. Prior to 1981 standards adopted by the Codex Alimentarius were published individually as Recommended Standards (CAC/RS series).

The Second Edition of the Codex Alimentarius is now being revised and updated to take into account decisions made by the 24th Session of the Codex Alimentarius Commission, July 2001.

INTRODUCTION

STATUTES OF THE CODEX ALIMENTARIUS COMMISSION AND IMPLEMENTATION OF THE FOOD STANDARDS PROGRAMME BY THE COMMISSION

The Codex Alimentarius Commission was established to implement the Joint FAO/WHO Food Standards Programme, the purpose of which is, as set down in the Statutes of the Commission, to protect the health of consumers and to ensure fair practices in the food trade; to promote coordination of all food standards work undertaken by international governmental and non-governmental organizations; to determine priorities and initiate and guide the preparation of draft standards through and with the aid of appropriate organizations; to finalize standards, and, after acceptance by governments, publish them in a Codex Alimentarius either as regional or world-wide standards[1].

The Statutes of the Codex Alimentarius Commission have been approved by the Governing bodies of the FAO and WHO. The Commission is a subsidiary body of these two parent Organizations. The Statutes and Rules of the Commission are to be found in the Procedural Manual of the Commission.

THE CODEX ALIMENTARIUS

Purpose

The Codex Alimentarius is a collection of internationally adopted food standards presented in a uniform manner. These food standards aim at protecting consumers' health and ensuring fair practices in the food trade. The Codex Alimentarius also includes provisions of an advisory nature in the form of codes of practice, guidelines and other recommended measures to assist in achieving the purposes of the Codex Alimentarius. The publication of the Codex Alimentarius is intended to guide and promote the elaboration and establishment of definitions and requirements for foods, to assist in their harmonization and, in doing so, to facilitate international trade.

Scope

The Codex Alimentarius includes standards for all the principal foods, whether processed, semi-processed or raw, for distribution to the consumer. Materials for further processing into foods are included to the extent necessary to achieve the purposes of the Codex Alimentarius as defined. The Codex Alimentarius includes provisions in respect of the hygienic and nutritional quality of food, including microbiological norms, provisions for food additives, pesticide residues, contaminants, labelling and presentation, and methods of analysis and sampling. It also includes provisions of an advisory nature in the form of codes of practice, guidelines and other recommended measures. Codex standards contain requirements for food aimed at ensuring for the consumer a sound, wholesome food product free from adulteration, correctly labelled and presented.

---

[1] The Codex Alimentarius Commission decided, at its 14th Session in July 1981, that its standards, which are sent to all Member States and Associate Members of FAO and/or WHO for acceptance, together with details of notifications received from governments with respect to the acceptance or otherwise of the standards and other relevant information, constitute the Codex Alimentarius.

Acceptance

The standards and maximum residue limits adopted by the Codex Alimentarius Commission are intended for formal acceptance by governments in accordance with its General Principles.

The standards and maximum limits for residues of pesticides and veterinary drugs in foods and feeds, accompanied by an appropriate communication, are sent for action to Ministries of Agriculture or Ministries of Foreign Affairs, as appropriate, of Member Nations of FAO and to Ministries of Health of Member States of WHO. The standards and maximum limits for pesticide residues and veterinary drugs, accompanied by the communication referred to, are also sent to national Codex Contact Points, FAO and WHO Regional Offices, FAO Representatives, Embassies in Rome and Permanent Missions to the UN in Geneva.

The standards and maximum limits for residues of pesticides and veterinary drugs in foods and feeds, which have taken a number of years to develop, are the product of a wide measure of international agreement and cooperation. They are compatible with the norms considered by FAO and WHO as best guaranteeing the protection of the health of consumers and the facilitation of international trade in food.

Current volume

This volume contains: the Codex Standards for Fish and Fishery Products and the Guidelines for the Sensory Evaluation of Fish and Shellfish in Laboratories.

The Codes of Practice for Fish and Fishery Products will be published in Volume 9B.

The Codex Alimentarius Commission at its 20th Session in 1993 adopted the Codex Standard for Dried Shark Fins.

The Codex Alimentarius Commission, at its 21st Session in 1995 adopted the Codex Standard for Quick Frozen Raw Squid, the General Standard for Quick Frozen Fish Fillets and the revised versions of all other standards for fish and fishery products.

The Codex Alimentarius Commission, at its 23rd Session in 1999 adopted the Guidelines for Sensory Evaluation of Fish and Shellfish in Laboratories.

The Standards for Fish and Fishery Products were previously contained in the Codex Alimentarius, First Edition, Volume V (1981).

TABLE OF CONTENTS

SECTION 1

QUICK FROZEN FISH AND FISHERY PRODUCTS

CODEX GENERAL STANDARD FOR QUICK FROZEN FISH FILLETS

CODEX STAN 190 - 1995

1.   SCOPE

This standard applies to quick frozen fillets of fish as defined below and offered for direct consumption without further processing. It does not apply to products indicated as intended for further processing or for other industrial purposes.

2.   DESCRIPTION

2.1   PRODUCT DEFINITION

Quick frozen fillets are slices of fish of irregular size and shape which are removed from the carcass of the same species of fish suitable for human consumption by cuts made parallel to the backbone and sections of such fillets cut so as to facilitate packing, and processed in accordance with the process definitions given in Section 2.2.

2.2   PROCESS DEFINITION

The product after any suitable preparation shall be subjected to a freezing process and shall comply with the conditions laid down hereafter. The freezing process shall be carried out in appropriate equipment in such a way that the range of temperature of maximum crystallization is passed quickly. The quick freezing process shall not be regarded as complete unless and until the product temperature has reached -18°C (0°F) or colder at the thermal centre after thermal stabilization. The product shall be kept deep frozen so as to maintain the quality during transportation, storage and distribution.

These products shall be processed and packaged so as to minimize dehydration and oxidation.

The recognized practice of repacking quick frozen products under controlled conditions which will maintain the quality of the product, followed by the reapplication of the quick freezing process as defined, is permitted.

2.3   PRESENTATION

2.3.1   Any presentation of the product shall be permitted provided that it:

(a)   meets all requirements of this standard, and

(b)   is adequately described on the label to avoid confusing or misleading the consumer.

2.3.2   Fillets may be presented as boneless, provided that boning has been completed including the removal of pin-bones.

3.   ESSENTIAL COMPOSITION AND QUALITY FACTORS

3.1   FISH

Quick frozen fish fillets shall be prepared from sound fish which are of a quality fit to be sold fresh for human consumption.

3.2   GLAZING

If glazed, the water used for glazing or preparing glazing solutions shall be of potable quality or shall be clean sea-water. Potable water is fresh-water fit for human consumption. Standards of potability shall not be less than those contained in the latest edition of the WHO "International Guidelines for Drinking Water Quality". Clean sea-water is sea-water which meets the same microbiological standards as potable water and is free from objectionable substances.

3.3   OTHER INGREDIENTS

All other ingredients used shall be of food grade quality and conform to all applicable Codex standards.

3.4   DECOMPOSITION

The products shall not contain more than 10 mg/100 g of histamine based on the average of the sample unit tested. This shall apply only to species of *Clupeidae, Scombridae, Scombresocidae, Pomatomidae* and *Coryphaenedae* families.

3.5   FINAL PRODUCT

Products shall meet the requirements of this standard when lots examined in accordance with Section 9 comply with the provisions set out in Section 8. Products shall be examined by the methods given in Section 7.

4.   FOOD ADDITIVES

Additive	Maximum level in the final product
Moisture/Water Retention Agents	
339(i) Monosodium orthophosphate	10 g/kg expressed as P_2O_5, singly or in combination (includes natural phosphate)
340(i) Monopotassium orthophosphate	
450(iii) Tetrasodium diphosphate	
450(v) Tetrapotassium diphosphate	
451(i) Pentasodium triphosphate	
451(ii) Pentapotassium triphosphate	
452(i) Sodium polyphosphate	
452(iv) Calcium, polyphosphates	
401 Sodium alginate	GMP
Antioxidants	
301 Sodium ascorbate	GMP
303 Potassium ascorbate	GMP

5.   HYGIENE AND HANDLING

5.1   The final product shall be free from any foreign material that poses a threat to human health.

5.2   When tested by appropriate methods of sampling and examination prescribed by the Codex Alimentarius Commission , the product:

(i)   shall be free from microorganisms or substances originating from microorganisms in amounts which may present a hazard to health in accordance with standards established by the Codex Alimentarius Commission;

(ii)   shall not contain histamine that exceeds 20 mg/100 g. This applies only to species of *Clupeidae, Scombridae, Scombresocidae, Pomatomidae* and *Coryphaenedae* families;

(iii)   shall not contain any other substance in amounts which may present a hazard to health in accordance with standards established by the Codex Alimentarius Commission.

5.3   It is recommended that the product covered by the provisions of this standard be prepared and handled in accordance with the appropriate sections of the Recommended International Code of Practice - General Principles of Food Hygiene (CAC/RCP 1-1969, Rev. 3-1997) and the following relevant Codes:

(i)   the Recommended International Code of Practice for Frozen Fish (CAC/RCP 16-1978);

(ii)   The Recommended International Code of Practice for the Processing and Handling of Quick Frozen Foods (CAC/RCP 8-1976).

(iii)   The sections on the Products of Aquaculture in the Proposed Draft International Code of Practice for Fish and Fishery Products (under elaboration)[1]

6.   LABELLING

In addition to the General Standard for the Labelling of Prepackaged Foods (CODEX STAN 1-1985, Rev.1-1991), the following specific provisions apply:

6.1   NAME OF THE FOOD

6.1.1   The name of the product as declared on the label shall be "... fillets" or "fillets of..." according to the law, custom or practice in the country in which the product is to be distributed.

6.1.2   There shall appear on the label reference to the form of presentation in close proximity to the name of the food in such additional words or phrases that will avoid misleading or confusing the consumer.

6.1.3   The term "quick frozen", shall also appear on the label, except that the term "frozen" may be applied in countries where this term is customarily used for describing the product processed in accordance with subsection 2.2 of this standard.

6.1.4   The label shall state that the product should be maintained under conditions that will maintain the quality during transportation, storage and distribution.

6.1.5   If the product has been glazed with sea-water, a statement to this effect shall be made.

6.2   NET CONTENTS (GLAZED PRODUCTS)

Where the food has been glazed the declaration of net contents of the food shall be exclusive of the glaze.

6.3   STORAGE INSTRUCTIONS

The label shall include terms to indicate that the product shall be stored at a temperature of -18° C or colder.

6.4   LABELLING OF NON-RETAIL CONTAINERS

Information on the above provisions shall be given either on the container or in accompanying documents, except that the name of the product, lot identification, and the name and address of the manufacturer or packer as well as storage instructions, shall appear on the container.

However, lot identification, and the name and address of the manufacturer or packer may be replaced by an identification mark provided that such a mark is clearly identifiable with the accompanying documents.

7.   SAMPLING, EXAMINATION AND ANALYSES

7.1   SAMPLING

(i)   Sampling of lots for examination of the product shall be in accordance with the FAO/WHO Codex Alimentarius Sampling Plans for Prepackaged Foods (AQL-6.5) (CODEX STAN 233-1969). A sample unit is the primary container or for individually quick frozen products is at least a 1 kg portion of the sample unit.

---

[1]   The Proposed Draft Code of Practice, when finalized, will replace all current Codes of Practice for Fish and Fishery Products

(ii)   Sampling of lots for examination of net weight shall be carried out in accordance with an appropriate sampling plan meeting the criteria established by the Codex Alimentarius Commission.

7.2   SENSORY AND PHYSICAL EXAMINATION

Samples taken for sensory and physical examination shall be assessed by persons trained in such examination and in accordance with procedures elaborated in Sections 7.3 through 7.6, Annex A and the *Guidelines for the Sensory Evaluation of Fish and Shellfish in Laboratories (CAC/GL 31 - 1999)*

7.3   DETERMINATION OF NET WEIGHT

7.3.1   The net weight (exclusive of packaging material) of each sample unit representing a lot shall be determined in the frozen state.

7.3.2   Determination of Net Weight of Products Covered by Glaze

As soon as the package is removed from low temperature storage, open immediately and place the contents under a gentle spray of cold water. Agitate carefully so that the product is not broken. Spray until all ice glaze that can be seen or felt is removed. Remove adhering water by the use of paper towel and weight the product in a tared pan.

7.4   PROCEDURE FOR THE DETECTION OF PARASITES (TYPE 1 METHOD) IN SKINLESS FILLETS

The entire sample unit is examined non-destructively by placing appropriate portions of the thawed sample unit on a 5 mm thick acryl sheet with 45% translucency and candled with a light source giving 1500 lux 30 cm above the sheet.

7.5   DETERMINATION OF GELATINOUS CONDITION

According to the AOAC Methods - "Moisture in Meat and Meat Products, Preparation of Sample Procedure"; 983.18 and "Moisture in Meat" (Method A); 950.46.

7.6   COOKING METHODS

The following procedures are based on heating the product to an internal temperature of 65 - 70°C. The product must not be overcooked. Cooking times vary according to the size of the product and the temperatures used. The exact times and conditions of cooking for the products should be determined by prior experimentation.

Baking Procedure: Wrap the product in aluminum foil and place it evenly on a flat cookie sheet or shallow flat pan.

Steaming Procedure: Wrap the product in aluminum foil and place it on a wire rack suspended over boiling water in a covered container.

Boil-in-Bag Procedure: Place the product in a boilable film-type pouch and seal. Immerse the pouch in boiling water and cook.

Microwave Procedure: Enclose the product in a container suitable for microwave cooking. If plastic bags are used, check to ensure that no odour is imparted from the plastic bags. Cook according to equipment instructions.

7.7   DETERMINATION OF HISTAMINE

AOAC 977.13.

8.   DEFINITION OF DEFECTIVES

A sample unit shall be considered as defective when it exhibits any of the properties defined below:

8.1   DEHYDRATION

Greater than 10% of the surface area of the sample unit or for pack sizes described below, exhibits excessive loss of moisture clearly shown as white or yellow abnormality on the surface, which masks the colour of the flesh and penetrates below the surface, and cannot be easily removed by scraping with a knife or other sharp instrument without unduly affecting the appearance of the product.

Pack Size	Defect Area
a) 200 g units	25 cm^2
b) 201 - 500 g units	50 cm^2
c) 501 - 5000 g units	150 cm^2

8.2   FOREIGN MATTER

The presence in the sample unit of any matter, which has not been derived from fish, does not pose a threat to human health, and is readily recognized without magnification or is present at a level determined by any method including magnification that indicates non-compliance with good manufacturing and sanitation practices.

8.3   PARASITES

The presence of two or more parasites per kg of the sample unit detected by the method described in 7.4 with a capsular diameter greater than 3 mm or a parasite not encapsulated and greater than 10 mm in length.

8.4   BONES (IN PACKS DESIGNATED BONELESS)

More than one bone per kg of product greater or equal to 10 mm in length, or greater or equal to 1 mm in diameter; a bone less than or equal to 5 mm in length, is not considered a defect if its diameter is not more than 2 mm. The foot of a bone (where it has been attached to the vertebra) shall be disregarded if its width is less than or equal to 2 mm, or if it can easily be stripped off with a fingernail.

8.5   ODOUR AND FLAVOUR

A sample unit affected by persistent and distinct objectionable odours or flavours characteristic of decomposition, rancidity or feed.

8.6   FLESH ABNORMALITIES

A sample unit affected by excessive gelatinous condition of the flesh together with greater than 86% moisture found in any individual fillet or a sample unit with pasty texture resulting from parasitic infestation affecting more than 5% of the sample unit by weight.

9.   LOT ACCEPTANCE

A lot will be considered as meeting the requirements of this standard when:

(i)   the total number of "defectives" as classified according to Section 8 does not exceed the acceptance number (c) of the appropriate sampling plan in the Sampling Plans for Prepackaged Foods (AQL-6.5) - (CODEX STAN 233-1969);

(ii)   the average net contents of all containers examined is not less than the declared weight, provided there is no unreasonable shortage in any containers;

(iii)   the Food Additives, Hygiene and Handling and the Labelling requirements of Sections 4, 5 and 6 are met.

"ANNEX A": SENSORY AND PHYSICAL EXAMINATION

1.   Complete net weight determination, according to defined procedures in Section 7.3 (de-glaze as required).

2.   Examine the frozen fillets for the presence of dehydration by measuring those areas which can only be removed with a knife or other sharp instrument. Measure the total surface area of the sample unit, and calculate the percentage affected.

3.   Thaw and individually examine each fillet in the sample unit for the presence of foreign matter, parasites, bone where applicable, odour, and flesh abnormality defects.

4.   In cases where a final decision on odour cannot be made in the thawed uncooked sate, a small portion of the disputed material (approximately 200 g) is sectioned from the sample unit and the odour and flavour confirmed without delay by using one of the cooking methods defined in Section 7.6.

5.   In cases where a final decision on gelatinous condition cannot be made in the thawed uncooked state, the disputed material is sectioned from the product and gelatinous condition confirmed by cooking as defined in Section 7.6 or by using the procedure in Section 7.5 to determine if greater than 86% moisture is present in any fillet. If a cooking evaluation is inconclusive, then the procedure in 7.5 would be used to make the exact determination of moisture content.

CODEX STANDARD FOR QUICK FROZEN BLOCKS OF FISH FILLET, MINCED FISH FLESH AND MIXTURES OF FILLETS AND MINCED FISH FLESH

CODEX STAN 165-1989, REV. 1 - 1995

1. SCOPE

This standard applies to quick frozen blocks of cohering fish flesh, prepared from fillets[1] or minced fish flesh or a mixture of fillets and minced fish flesh, which are intended for further processing.

2. DESCRIPTION

2.1 PRODUCT DEFINITION

Quick frozen blocks are rectangular or other uniformly shaped masses of cohering fish fillets, minced fish or a mixture thereof, which are suitable for human consumption, comprising:

(i) a single species; or

(ii) a mixture of species with similar sensory characteristics.

2.1.1 Fillets are slices of fish of irregular size and shape which are removed from the carcass by cuts made parallel to the back bone and pieces of such fillets, with or without the skin.

2.1.2 Minced fish flesh used in the manufacture of blocks are particles of skeletal muscle which have been separated from and are essentially free from bones, viscera and skin.

2.2 PROCESS DEFINITION

The product after any suitable preparation shall be subjected to a freezing process and shall comply with the conditions laid down hereafter. The freezing process shall be carried out in appropriate equipment in such a way that the range of temperature of maximum crystallization is passed quickly. The quick freezing process shall not be regarded as complete unless and until the product temperature has reached -18°C or colder at the thermal centre after thermal stabilization. The product shall be kept deep frozen so as to maintain the quality during transportation, storage and distribution.

Industrial repacking or further processing of intermediate quick frozen material under controlled conditions which maintain the quality of the product followed by the reapplication of the quick freezing process is permitted.

These products shall be processed and packaged so as to minimize dehydration and oxidation.

2.3 PRESENTATION

Any presentation of the product shall be permitted provided that it:

2.3.1 meets all requirements of this standard, and

2.3.2 is adequately described on the label to avoid confusing or misleading the consumer.

2.3.3 Blocks may be presented as boneless, provided that boning has been completed including the removal of pin-bones.

---

[1] Including pieces of fillets.

3.   ESSENTIAL COMPOSITION AND QUALITY FACTORS

3.1   FISH

Quick frozen blocks shall be prepared from fillets or minced flesh of sound fish which are of a quality fit to be sold fresh for human consumption.

3.2   GLAZING

If glazed, the water used for glazing or preparing glazing solutions shall be of potable quality or shall be clean sea-water. Potable water is fresh-water fit for human consumption. Standards of potability shall not be less than those contained in the latest edition of the WHO "International Guidelines for Drinking Water Quality". Clean sea-water is sea-water which meets the same microbiological standards as potable water and is free from objectionable substances.

3.3   OTHER INGREDIENTS

All other ingredients used shall be of food grade quality and conform to all applicable Codex standards.

3.4   DECOMPOSITION

The products shall not contain more than 10 mg/100 g of histamine based on the average of the sample unit tested. This shall apply only to species of *Clupeidae, Scombridae, Scombresocidae, Pomatomidae* and *Coryphaenedae* families.

3.5   FINAL PRODUCT

Products shall meet the reguirements of this standard when lots examined in accordance with Section 9 comply with the provisions set out in Section 8. Products shall be examined by the methods given in Section 7.

4.   FOOD ADDITIVES

Only the use of the following additives is permitted.

Additive	Maximum Level in the Final Product
Moisture/Water Retention Agents	
339(i) Monosodium orthophosphate	10 g/kg expressed as
340(i) Monopotassium orthophosphate	P_2O_5, singly or in
450(iii) Tetrasodium diphosphate	combination (includes
450(v) Tetrapotassium diphosphate	natural phosphate)
451(i) Pentasodium triphosphate	
451(ii) Pentapotassium triphosphate	
452(i) Sodium polyphosphate	
452(iv) Calcium, polyphosphates	
401 Sodium alginate	GMP
Antioxidants	
300 Ascorbic acid	GMP
301 Sodium ascorbate	
303 Potassium ascorbate	
304 Ascorbyl palmitate	1 g/kg

In Minced Fish Flesh Only

Acidity Regulator

330	Citric acid	GMP
331	Sodium citrate	
332	Potassium citrate	

Thickeners

412	Guar gum	GMP
410	Carob bean (Locust bean) gum	
440	Pectins	
466	Sodium carboxymethyl cellulose	
415	Xanthan gum	
407	Carrageenan and its Na, K, NH$_4$ salts (including Furcelleran)	
407a	Processed *Euchema* Seaweed (PES)	
461	Methyl cellulose	

5. HYGIENE AND HANDLING

5.1 The final product shall be free from any foreign material that poses a threat to human health.

5.2 When tested by appropriate methods of sampling and examination prescribed by the Codex Alimentarius Commission, the product:

(i) shall be free from microorganisms or substances originating from microorganisms in amounts which may represent a hazard to health in accordance with standards established by the Codex Alimentarius Commission;

(ii) shall not contain histamine that exceeds 20 mg/100 g in any sample unit. This applies only to species of Clupeidae, Scombridae, Scombresocidae, Pomatomidae and Coryphaenedae families;

(iii) shall not contain any other substances in amounts which may represent a hazard to health in accordance with standards established by the Codex Alimentarius Commission.

5.3 It is recommended that the product covered by the provisions of this standard be prepared and handled in accordance with the appropriate sections of the Recommended International Code of Practice - General Principles of Food Hygiene (CAC/RCP 1-1969, Rev. 3-1997) and the following relevant Codes:

(i) The Recommended International Code of Practice for Frozen Fish (CAC/RCP 16-1978);

(ii) The Recommended International Code of Practice for Frozen Battered and/or Breaded Fishery Products (CAC/RCP 35-1985);

(iii) The Recommended International Code of Practice for Minced Fish Prepared by Mechanical Separation (CAC/RCP 27-1983).

(iv)   The Recommended International Code of Practice for the Processing and Handling of Quick Frozen Foods (CAC/RCP 8-1976).

(v)   The sections on the Products of Aquaculture in the Proposed Draft International Code of Practice for Fish and Fishery Products (under elaboration)[2]

6.   LABELLING

In addition to the provisions of the Codex General Standard for the Labelling of Prepackaged Foods (CODEX STAN 1-1985, Rev. 1-1991) the following specific provisions apply;

6.1   THE NAME OF THE FOOD

6.1.1   The name of the food shall be declared as "x y blocks" in accordance with the law, custom or practice of the country in which the product is distributed, where "x" shall represent the common name(s) of the species packed and "y" shall represent the form of presentation of the block (see Section 2.3).

6.1.2   If the product has been glazed with sea-water, at statement to this effect shall be made

6.1.3   The name "quick frozen", shall also appear on the label, except that the term "frozen" may be applied in countries where this term is customarily used for describing the product processed in accordance with subsection 2.2 of this standard.

6.1.4   The proportion of mince in excess of 10% of net fish content shall be declared stating the percentage ranges: 10-25, >25-35, etc. Blocks with more than 90% mince are regarded as mince blocks.

6.1.5   The label shall state that the product should be maintained under conditions that will maintain the quality during transportation, storage and distribution.

6.2   NET CONTENTS (GLAZED BLOCKS)

Where the food has been glazed, the declaration of net contents of the food shall be exclusive of the glaze.

6.3   STORAGE INSTRUCTIONS

The label shall include terms to indicate that the product shall be stored at a temperature of -18°C or colder.

6.4   LABELLING OF NON-RETAIL CONTAINERS

Information specified above shall be given either on the container or in accompanying documents, except that the name of the product, lot identification, and the name and address of the manufacturer or packer as well as storage instructions, shall appear on the container.

However, lot identification, and the name and address of the manufacturer or packer may be replaced by an identification mark provided that such mark is clearly identifiable with the accompanying documents.

7.   SAMPLING, EXAMINATION AND ANALYSES

7.1   SAMPLING PLAN FOR FISH BLOCKS

(i)   Sampling of lots for examination of the product shall be in accordance with the sampling plan defined below. The sample unit is the entire block.

---

[2]   The Proposed Draft Code of Practice, when finalized, will replace all current Codes of Practice for Fish and Fishery Products

Lot Size (Number of blocks)	Sample Size (Number of blocks to be tested, n)	Acceptance number (c)
< 15	2	0
16 - 50	3	0
51 – 150	5	1
151 - 500	8	1
501 – 3200	13	2
3201 – 35000	20	3
> 35000	32	5

If the number of defective blocks in the sample is less than or equal to c, accept the lot; otherwise, reject the lot.

(ii)   Sampling of lots for examination of net weight shall be carried out in accordance with an appropriate sampling plan meeting the established criteria established by the Codex Alimentarius Commission.

7.2   SENSORY AND PHYSICAL EXAMINATION

Samples taken for sensory and physical examination shall be assessed by persons trained in such examination and in accordance with procedures elaborated in Sections 7.3 through 7.7 and Annex A and in accordance with the Guidelines for the Sensory Evaluation of Fish and Shellfish (CAC/GL 31-1999).

7.3   DETERMINATION OF NET WEIGHT

7.3.1   Determination of Net Weight of Product Not Covered by Glaze

The net weight (exclusive of packaging material) of each sample unit representing a lot shall be determined in the frozen state.

7.3.2   Determination of Net Weight of Products Covered by Glaze

As soon as the package is removed from frozen temperature storage, open immediately and place the contents under a gentle spray of cold water until all ice glaze that can be seen or felt is removed. Remove adhering water by the use of paper towel and weigh the product.

An alternate method is outlined in Annex B.

7.4   PROCEDURE FOR THE DETECTION OF PARASITES FOR SKINLESS BLOCKS OF FISH FILLETS (TYPE I METHOD)

The entire sample unit is examined non-destructively by placing appropriate portions of the thawed sample unit on a 5 mm thick acryl sheet with 45% translucency and candled with a light source giving 1500 lux 30 cm above the sheet.

7.5 DETERMINATION OF PROPORTIONS OF FILLET AND MINCED FISH IN QUICK FROZEN BLOCKS PREPARED FROM MIXTURES OF FILLETS AND MINCED FISH [3] [4]

According to the AOAC Method 988.09.

7.6 DETERMINATION OF GELATINOUS CONDITION

According to the AOAC Methods - "Moisture in Meat and Meat Products, Preparation of Sample Procedure"; AOAC 983.18 and "Moisture in Meat" Method A, 950.46.

7.7 COOKING METHODS

The following procedures are based on heating the product to an internal temperature of 65 -70°C. The product must not be overcooked. Cooking times vary according to the size of the product and the temperatures used. The exact times and conditions of cooking for the products should be determined by prior experimentation.

Baking Procedure: Wrap the product in aluminum foil and place it evenly on a flat cookie sheet or shallow flat pan.

Steaming Procedure: Wrap the product in aluminum foil and place it on a wire rack suspended over boiling water in a covered container.

Boil-In-Bag Procedure: Place the product into a boilable film-type pouch and seal. Immerse the pouch into boiling water and cook.

Microwave Procedure: Enclose the product in a container suitable for microwave cooking. If plastic bags are used, check to ensure that no odour is imparted from the plastic bags. Cook according to equipment instructions.

7.8. THAWING PROCEDURE FOR QUICK FROZEN BLOCKS

Air Thaw Method:

Frozen fish blocks are removed from the packaging. The frozen fish blocks are individually placed into snug fitting impermeable plastic bags or a humidity controlled environment with a relative humidity of at least 80%. Remove as much air as possible from the bags and seal. The frozen fish blocks sealed in plastic bags are placed on individual trays and thawed at air temperature of 25°C (77°F) or lower. Thawing is completed when the product can be readily separated without tearing. Internal block temperature should not exceed 7°C (44.6°F).

Water Immersion Method:

Frozen fish blocks are removed from the packaging. The frozen fish blocks are sealed in plastic bags. Remove as much air as possible from the bags and seal. The frozen fish blocks are placed into a circulating water bath with temperatures maintained at 21°C $\pm$ 1.5°C (70°F $\pm$ 3°F). Thawing is completed when the product can be easily separated without tearing. Internal block temperature should not exceed 7°C (44.6°F).

7.9 DETERMINATION OF HISTAMINE

AOAC 977.13.

8. DEFINITION OF DEFECTIVES

The sample unit shall be considered defective when it exhibit any of the properties defined below.

---

[3]   This method has been evaluated for cod only but, in principle, should be appropriate to other fish species or mixed species.

[4]   This method is accurate for levels of mince greater than 10%.

Codex Alimentarius
Volume 9A - 2001

\- 15 -

Quick Frozen Blocks of Fish Fillet,
Minced Fish Flesh and Mixtures
CODEX STAN 165-1989, Rev. 1-1995

8.1 DEEP DEHYDRATION

Greater than 10% of the surface area of the sample unit exhibits excessive loss of moisture clearly shown as white or yellow abnormality on the surface which masks the colour of the flesh and penetrates below the surface, and cannot be easily removed by scraping with a knife or other sharp instrument without unduly affecting the appearance of the block.

8.2 FOREIGN MATTER

The presence in the sample unit of any matter which has not been derived from fish (excluding packing material), does not pose a threat to human health, and is readily recognized without magnification or is present at a level determined by any method including magnification that indicates non-compliance with good manufacturing and sanitation practices.

8.3 PARASITES

The presence of two or more parasites per kg of the sample unit detected by a method described in 7.4 with a capsular diameter greater than 3 mm or a parasite not encapsulated and greater than 10 mm in length.

8.4 BONES (IN PACKS DESIGNATED BONELESS)

More than one bone per kg of product greater or equal to 10mm in length, or greater or equal to 1 mm in diameter; a bone less than or equal to 5 mm in length, is not considered a defect if its diameter is not more than 2 mm. The foot of a bone (where it has been attached to the vertebra) shall be disregarded if its width is less than or equal to 2 mm, or if it can easily be stripped off with a fingernail.

8.5 ODOUR AND FLAVOUR

A sample unit affected by persistent and distinct objectionable odours or flavours indicative of decomposition or rancidity or of feed.

8.6 FLESH ABNORMALITIES

A sample unit affected by excessive gelatinous condition of the flesh together with greater than 86% moisture found in any individual fillet, or a sample unit with pasty texture resulting from parasitic infestation affecting more than 5% of the sample unit by weight.

9. LOT ACCEPTANCE

A lot shall be considered as meeting the requirements of this standard when:

(i)   the total number of defective sample units as classified according to Section 8 does not exceed the acceptance number (c) of the sampling plan in Section 7; and

(ii)   the average net weight of all sample units is not less than the declared weight, provided there is no unreasonable shortage in any container; and

(iii)   the Food Additives, Hygiene and Labelling requirements of Sections 4, 5 and 6 are met.

"ANNEX A": SENSORY AND PHYSICAL EXAMINATION

1. Complete net weight determination, according to defined procedures in Section 7.3 (de-glaze as required).

2. Examine the frozen block for the presence of dehydration by measuring those areas which can only be removed with a knife or other sharp instrument. Measure the total surface area of the sample unit, and calculate the percentage affected.

3. Thaw and individually examine each block in the sample unit for the presence of foreign matter, bone where applicable, odour, and textural defects.

4. In cases where a final decision on odour can not be made in the thawed uncooked sate, a small portion of the disputed material (approximately 200 g) is sectioned from the block and the odour and flavour confirmed without delay by using one of the cooking methods defined in Section 7.8.

5. In cases where a final decision on gelatinous condition cannot be made in the thawed uncooked state, the disputed material is sectioned from the block and the gelatinous condition confirmed by cooking as defined in Section 7.7. or by using procedure in Section 7.6. to determine if greater than 86% moisture is present in any fillet. If cooking evaluation is inconclusive, then procedure in 7.6. would be used to make the exact determination of moisture content.

ANNEX B: METHOD FOR THE DETERMINATION OF NET CONTENT OF FROZEN FISH BLOCKS COVERED BY GLAZE

Glazing is not used for Q.F. blocks of white fish. Only Q.F. blocks of herring, mackerel and other brown (fat) fish are glazed, which are destined for further processing (canning, smoking). For such blocks the following procedure may be applicable (tested with block frozen shrimps).

1. **PRINCIPLE:**

The pre-weighed glazed sample is immersed into a water bath by hand till all glaze is removed (as felt by fingers). As soon as the surface becomes rough, the still frozen sample is removed from the water bath and dried by use of a paper towel before estimating the net product content by repeated weighing. By this procedure thaw drip losses and/or re-freezing of adhering moisture can be avoided.

2. **EQUIPMENT:**
   - Balance - sensitive to 1 g
   - Water bath, preferably with adjustable temperature
   - Circular sieve with a diameter of 20 cm and 1-3 mm mesh apertures (ISO R 565)
   - Paper or cloth towels with smooth surface
   - A freezed box should be available at the working place

3. **PREPARATION OF SAMPLES AND WATER BATH:**
   - The product temperature should be adjusted to -18/-20°C to achieve standard deglazing conditions (especially necessary if a standard deglazing period shall be defined in case of regular shaped products).
   - After sampling from the low temperature store remove, if present, external ice crystals or snow from the package with the frozen product.
   - The water bath shall contain an amount of fresh potable water equal to about 10 times of the declared weight of the product; the temperature should be adjusted on about 15°C to 35°C.

4. **DETERMINATION OF GROSS-WEIGHT "A":**

After removal of the package, the weight of the glazed product is determined: In case of single fish fillets, single weights are recorded (A 1-A n). The weighed samples are placed intermediately into the freezer box.

5. **REMOVAL OF GLAZE:**

The pre-weighed samples/sub-samples are transferred into the water bath and kept immersed by hand. The product may be carefully agitated, till no more glaze can be felt by the finger-tips on the surface of the product: change from slippery to rough. Needed time, depending on size/shape and glaze content of the product, 10 to 60 sec. (and more in case of higher glaze contents or if frozen together).

For block-frozen products in consumer packs (also for single glaze products, which are frozen together during storage) the following (preliminary) procedure may be applicable: The pre-weighed block or portion is transferred onto a suitable sized sieve and immersed into the water bath. By slight pressure of the fingers separating deglazed portions are removed fractionally. Short immersing is repeated, if glaze residues are still present.

6. **DETERMINATION OF NET WEIGHT "B"**

The deglazed sample/sub-sample, after removal of adhering water by use of a towel (without pressure) is immediately weighed. Single net-weights of sub-samples are summed up: B_{1-n}.

Quick Frozen Blocks of Fish Fillet,   - 18 -   Codex Alimentarius
Minced Fish Flesh and Mixtures   *Volume 9A - 2001*
CODEX STAN 165-1989, Rev. 1-1995

1. **DETERMINATION OF GLAZE-WEIGHT "C"**

   Gross weight "A" – Net weight "B" = Glaze weight "C"

2. **CALCULATION OF PERCENTAGE PROPORTIONS:**

   % net content of the product $\quad "F" = \dfrac{"B"}{"A"} \times 100$

   % glaze – related to the gross weight of the product $\quad "G" = \dfrac{"C"}{"A"} \times 100$

   % glaze – related to the net weight of the product $\quad "H" = \dfrac{"C"}{"B"} \times 100$

CODEX STANDARD FOR QUICK FROZEN FINFISH, UNEVISCERATED AND EVISCERATED

CODEX STAN 36 - 1981, REV. 1 - 1995

1. SCOPE

This standard shall apply to frozen finfish uneviscerated and eviscerated[1].

2. DESCRIPTION

2.1 PRODUCT DEFINITION

Frozen finfish suitable for human consumption, with or without the head, from which the viscera or other organs may have been completely or partially removed.

2.2 PROCESS DEFINITION

The product, after any suitable preparation, shall be subjected to a freezing process and shall comply with the conditions laid down hereafter. The freezing process shall be carried out in appropriate equipment in such a way that the range of temperature of maximum crystallization is passed quickly. The quick freezing process shall not be regarded as complete unless and until the product temperature has reached -18°C or colder at the thermal centre after thermal stabilization. The product shall be kept deep frozen so as to maintain the quality during transportation, storage and distribution.

Industrial repacking of quick frozen products under controlled conditions which maintain the quality of the products followed by the reapplication of the quick freezing process is permitted.

Quick frozen finfish, shall be processed and packaged so as to minimize dehydration and oxidation.

2.3 PRESENTATION

Any presentation of the product shall be permitted provided that it:

2.3.1 meets all requirements of this standard; and

2.3.2 is adequately described on the label to avoid confusing or misleading the consumer.

3. ESSENTIAL COMPOSITION AND QUALITY FACTORS

3.1 FISH

Quick frozen finfish shall be prepared from sound fish which are of a quality fit to be sold fresh for human consumption.

3.2 GLAZING

If glazed, the water used for glazing or preparing glazing solutions shall be of potable quality or shall be clean sea-water. Potable water is fresh-water fit for human consumption. Standards of potability

---

[1]   It does not apply to fish frozen in brine intended for further processing.

shall not be less than those contained in the latest edition of the WHO "International Guidelines for Drinking Water Quality". Clean sea-water is sea-water which meets the same microbiological standards as potable water and is free from objectionable substances.

3.3 OTHER INGREDIENTS

All other ingredients used shall be of food grade quality and conform to all applicable Codex and WHO standards.

3.4 DECOMPOSITION

The products shall not contain more than 10 mg/100 g of histamine based on the average of the sample unit tested. This shall apply only to species of *Clupeidae, Scombridae, Scombresocidae, Pomatomidae* and *Coryphaenedae* families.

3.5 FINAL PRODUCT

Products shall meet the requirements of this standard when lots examined in accordance with Section 9 comply with the provisions set out in Section 8. Products shall be examined by the methods given in Section 7.

4. FOOD ADDITIVES

Only the use of the following additives is permitted.

Additive	Maximum Level in the Final Product
Antioxidants	
300 Ascorbic acid	GMP
301 Sodium ascorbate	
303 Potassium ascorbate	

5. HYGIENE AND HANDLING

5.1 The final product shall be free from any foreign material that poses a threat to human health.

5.2 When tested by appropriate methods of sampling and examination prescribed by the Codex Alimentarius Commission, the product:

(i) shall be free from microorganisms or substances originating from microorganisms in amounts which may present a hazard to health in accordance with standards established by the Codex Alimentarius Commission;

(ii) shall not contain histamine that exceeds 20 mg/100 g. This applies only to species of *Clupeidae, Scombridae, Scombresocidae, Pomatomidae* and *Coryphaenedae* families

(iii) shall not contain any other substance in amounts which may present a hazard to health in accordance with standards established by the Codex Alimentarius Commission.

5.3 It is recommended that the product covered by the provisions of this standard be prepared and handled in accordance with the appropriate sections of the Recommended International Code of Practice - General Principles of Food Hygiene (CAC/RCP 1-1969, Rev. 3-1997) and the following relevant Codes:

(i) the Recommended International Code of Practice for Frozen Fish (CAC/RCP 16-1978);

(ii) The Recommended International Code of Practice for the Processing and Handling of Quick Frozen Foods (CAC/RCP 8-1976).

(iii) The sections on the Products of Aquaculture in the Proposed Draft International Code of Practice for Fish and Fishery Products (under elaboration)[1]

6. LABELLING

In addition to the provisions of the Codex General Standard for the Labelling of Prepackaged Foods (CODEX STAN 1-1985, Rev. 1-1991) the following specific provisions apply:

6.1 THE NAME OF THE FOOD

6.1.1 In addition to the common or usual name of the species, the label, in the case of eviscerated fish, shall include terms indicating that the fish has been eviscerated and whether presented as "head-on" or "headless".

6.1.2 If the product has been glazed with sea-water, a statement to this effect shall be made.

6.1.3 The term "quick frozen", shall also appear on the label, except that the term "frozen" may be applied in countries where this term is customarily used for describing the product processed in accordance with subsection 2.2 of this standard.

6.1.4 The label shall state that the product should be maintained under conditions that will maintain the quality during transportation, storage and distribution.

6.2 NET CONTENTS (GLAZED PRODUCTS)

Where the food has been glazed the declaration of net contents of the food shall be exclusive of the glaze.

6.3 STORAGE INSTRUCTIONS

The label shall include terms to indicate that the product shall be stored at a temperature of -18°C or colder.

6.4 LABELLING OF NON-RETAIL CONTAINERS

Information specified above shall be given either on the container or in accompanying documents, except that the name of the food, lot identification, and the name and address, as well as storage instructions shall always appear on the container.

However, lot identification, and the name and address may be replaced by an identification mark, provided that such a mark is clearly identifiable with the accompanying documents.

7. SAMPLING, EXAMINATION AND ANALYSES

7.1 SAMPLING

(i) Sampling of lots for examination of the product shall be in accordance with the FAO/WHO Codex Alimentarius Sampling Plans for Prepackaged Foods (AQL-6.5) (CODEX STAN 233-1969). A sample unit is the individual fish or the primary container.

(ii) Sampling of lots for examination of net weight shall be carried out in accordance with an appropriate sampling plan meeting the criteria established by the Codex Alimentarius Commission.

7.2 SENSORY AND PHYSICAL EXAMINATION

Samples taken for sensory and physical examination shall be assessed by persons trained in such examination and in accordance with procedures elaborated in Sections 7.3, 7.4 and 7.5, Annex A and the

---

[1]   The Proposed Draft Code of Practice, when finalized, will replace all current Codes of Practice for Fish and Fishery Products

Guidelines for the Sensory Evaluation of Fish and Shellfish in Laboratories (CAC/GL 31 - 1999)

7.3 DETERMINATION OF NET WEIGHT

7.3.1 Determination of Net Weight of Products not Covered by Glaze

The net weight (exclusive of packaging material) of each sample unit representing a lot shall be determined in the frozen state.

7.3.2 Determination of Net Weight of Products Covered by Glaze

(To be elaborated).

7.4 THAWING

(To be elaborated).

7.5. DETERMINATION OF GELATINOUS CONDITIONS

According to the AOAC Methods- "Moisture in Meat and Meat Products, Preparation of Sample Procedure"; 883.18 and "Moisture in Meat" (Method A); 950.46.

7.6 COOKING METHODS

The following procedures are based on heating the product to an internal temperature of 65-70°C. The product must not be overcooked. Cooking times vary according to the size of the product and the temperatures used. The exact times and conditions of cooking for the product should be determined by prior experimentation.

Baking Procedure: Wrap the product in aluminum foil and place it evenly on a flat cookie sheet or shallow flat pan.

Steaming Procedure: Wrap the product in aluminum foil and place it on a wire rack suspended over boiling water in a covered container.

Boil-In-Bag Procedure: Place the product into a boilable film-type pouch and seal. Immerse the pouch into boiling water and cook.

Microwave Procedure: Enclose the product in a container suitable for microwave cooking. If plastic bags are used, check to ensure that no odour is imparted from the plastic bags. Cook according to equipment specifications.

7.7 DETERMINATION OF HISTAMINE

AOAC 977.13

8. DEFINITION OF DEFECTIVES

The sample unit shall be considered defective when it exhibits any of the properties defined below:

8.1 DEEP DEHYDRATION

Greater than 10% of the surface area of the block or greater than 10% of the weight of fish in the sample unit exhibits excessive loss of moisture clearly shown as white or yellow abnormality on the surface which masks the colour of the flesh and penetrates below the surface, and cannot be easily removed by scraping with a knife or other sharp instrument without unduly affecting the appearance of the fish.

8.2 FOREIGN MATTER

The presence in the sample unit of any matter which has not been derived from fish (excluding packaging material), does not pose a threat to human health, and is readily recognized without magnification or is present at a level determined by any method including magnification, that indicates non-compliance with good manufacturing and sanitation practices.

8.3 ODOUR AND FLAVOUR

A sample unit affected by persistent and distinct objectionable odours or flavours indicative of decomposition or of feed.

8.4. TEXTURE

8.4.1 Textural breakdown of the flesh, indicative of decomposition characterized by muscle structure which is mushy or paste-like, or by separation of flesh from the bones.

8.4.2 Flesh abnormalities

A sample unit affected by excessive gelatinous condition of the flesh together with greater then 86% moisture found in any individual fish or sample unit with pasty texture resulting from parasitic infestation affecting more than 5% of the sample unit by weight.

8.5 BELLY BURST

The presence of ruptured bellies in uneviscerated fish, indicative of decomposition.

9. LOT ACCEPTANCE

A lot shall be considered as meeting the requirements of this standard when:

(i) the total number of defectives as classified according to Section 8 does not exceed the acceptance number (c) of the appropriate sampling plan in the Sampling Plans for Prepackaged Foods (AQL-6.5) (CODEX STAN 233-1969);

(ii) the average net weight of all sample units is not less than the declared weight, provided there is no unreasonable shortage in any container; and

(iii) the Food Additives, Hygiene and Labelling requirements of Sections 4, 5 and 6 are met.

"ANNEX A": SENSORY AND PHYSICAL EXAMINATION

1. Complete net weight determination, according to defined procedures in Section 7.3 (de-glaze as required).

2. Examine the frozen sample unit for the presence of deep dehydration by measuring those areas or counting instances which can only be removed with a knife or other sharp instrument. Measure the total surface area of the sample unit, and calculate the percentage affected.

3. Thaw and individually examine each fish in the sample unit for the presence of foreign matter.

4. Examine each fish using the criteria outlined in Section 8. Flesh odours are examined by tearing or making a cut across the back of the neck such that the exposed surface of the flesh can be evaluated.

5. In cases where a final decision regarding the odour or texture can not be made in the thawed uncooked state, a small portion of the flesh (approximately 200 g) is sectioned from the product and the odour, flavour or texture confirmed without delay by using one of the cooking methods defined in Section 7.6.

6. In cases where a final decision on gelatinous condition cannot be made in the thawed uncooked state, the disputed material is sectioned from the product and gelatinous condition confirmed by cooking as defined in Section 7.6 or by using the procedure in Section 7.5 to determine if greater than 86% moisture is present in any fish. If a cooking evaluation is inconclusive, then the procedure in 7.5 would be used to make the exact determination of moisture content.

CODEX STANDARD FOR QUICK FROZEN FISH STICKS (FISH FINGERS), FISH PORTIONS AND FISH FILLETS - BREADED OR IN BATTER

CODEX STAN 166 - 1989, REV 1 - 1995

1. SCOPE

This standard applies to quick frozen fish sticks (fish fingers) and fish portions cut from quick frozen fish flesh blocks, or formed from fish flesh, and to natural fish fillets, breaded or batter coatings, singly or in combination, raw or partially cooked and offered for direct human consumption without further industrial processing.

2. DESCRIPTION

2.1 PRODUCT DEFINITION

2.1.1 A fish stick (fish finger) is the product including the coating weighing not less than 20 g and not more than 50 g shaped so that the length is not less than three times the greatest width. Each stick shall be not less than 10 mm thick.

2.1.2 A fish portion including the coating, other than products under 2.1.1, may be of any shape, weight or size.

2.1.3 Fish sticks or portions may be prepared from a single species of fish or from a mixture of species with similar sensory properties.

2.1.4 Fillets are slices of fish of irregular size and shape which are removed from the carcass by cuts made parallel to the back bone and pieces of such fillets, with or without the skin.

2.2 PROCESS DEFINITION

The product after any suitable preparation shall be subjected to a freezing process and shall comply with the conditions laid down hereafter. The freezing process shall be carried out in appropriate equipment in such a way that the range of temperature of maximum crystallization is passed quickly. The quick freezing process shall not be regarded as complete unless and until the product temperature has reached -18°C or colder at the thermal centre after thermal stabilization. The product shall be kept deep frozen so as to maintain the quality during transportation, storage and distribution.

Industrial repacking or further industrial processing of intermediate quick frozen material under controlled conditions which maintains the quality of the product, followed by the re-application of the quick freezing process, is permitted.

2.3 PRESENTATION

Any presentation of the product shall be permitted provided that it:

2.3.1 meets all the requirements of the standard, and

2.3.2 is adequately described on the label to avoid confusing or misleading the consumer.

3. ESSENTIAL COMPOSITION AND QUALITY FACTORS

3.1 RAW MATERIAL

3.1.1 Fish

Quick frozen breaded or battered fish sticks (fish fingers) breaded or battered fish portions and breaded or battered fillets shall be prepared from fish fillets or minced fish flesh, or mixtures thereof, of edible species which are of a quality such as to be sold fresh for human consumption.

3.1.2   Coating

The coating and all ingredients used therein shall be of food grade quality and conform to all applicable Codex standards.

3.1.3   Frying fat (oil)

A fat (oil) used in the cooking operation shall be suitable for human consumption and for the desired final product characteristic (see also Section 4).

3.2   FINAL PRODUCT

Products shall meet the requirements of this standard when lots examined in accordance with Section 9 comply with the provisions set out in Section 8. Products shall be examined by the methods given in Section 7.

3.3   DECOMPOSITION

The products shall not contain more than 10 mg/100 g of histamine based on the average of the sample unit tested. This shall apply only to species of *Clupeidae*, *Scombridae*, *Scombresocidae*, *Pomatomidae* and *Coryphaenedae* families.

4.   FOOD ADDITIVES

Only the use of the following additives is permitted.

Additive	Maximum level in the final product

For Fish Fillets and Minced Fish Flesh Only

Moisture/Water Retention Agents

		Maximum level
339(i)	Monosodium orthophosphate	10 g/kg expressed as P_2O_5, singly or in combination (includes natural phosphate)
340(i)	Monopotassium orthophosphate	
450(iii)	Tetrasodium diphosphate	
450(v)	Tetrapotassium diphosphate	
451(i)	Pentasodium triphosphate	
451(ii)	Pentapotassium triphosphate	
452(i)	Sodium polyphosphate	
452(iv)	Calcium, polyphosphates	
401	Sodium alginate	GMP

Antioxidants

300	Ascorbic acid	GMP
301	Sodium ascorbate	
303	Potassium ascorbate	
304	Ascorbyl palmitate	1 g/kg

In Addition, for Minced Fish Flesh Only

Acidity Regulator

330	Citric acid	GMP
331	Sodium citrate	
332	Potassium citrate	

Additive	Maximum level in the final product

Thickeners

412	Guar gum	GMP
410	Carob bean (Locust bean) gum	
440	Pectins	
466	Sodium carboxymethyl cellulose	
415	Xanthan gum	
407	Carrageenan and its Na, K, NH_4 salts (including Furcelleran)	
407a	Processed Euchema seaweed (PES)	
461	Methyl cellulose	

Food Additives for Breaded or Batter Coatings

Leavening Agents

341(i)	Monocalcium orthophosphate	1 g/kg expressed as P_2O_5, singly or in combination
341(ii)	Dicalcium orthophosphate	
541	Sodium aluminium phosphate, basic and acidic	
500	Sodium carbonates	GMP
501	Potassium carbonates	
503	Ammonium carbonates	

Flavour Enhancers

621	Monosodium glutamate	GMP
622	Monopotassium glutamate	

Colours

160b	Annatto extracts	20 mg/kg expressed as bixin
150a	Caramel I (plain)	GMP
160a(i)	β-carotene (Synthetic)	100 mg/kg singly or in combination
160e	β-apo-carotenal	

Thickeners

412	Guar gum	GMP
410	Carob bean (Locust bean) gum	
440	Pectins	
466	Sodium carboxymethyl cellulose	
415	Xanthan gum	
407	Carrageenan and its Na, K, NH_4 salts (including Furcelleran)	
407a	Processed *Euchema* Seaweed (PES)	
461	Methyl cellulose	
401	Sodium alginate	
463	Hydroxypropyl cellulose	
464	Hydroxypropyl methylcellulose	
465	Methylethylcellulose	

Emulsifiers

471	Monoglycerides of fatty acids	GMP
322	Lecithins	

Modified Starches

1401	Acid treated starches	GMP
1402	Alkaline treated starches	
1404	Oxidized starches	
1410	Monostarch phosphate	
1412	Distarch phosphate esterified with sodium trimetaphosphate; esterified with phosphorus oxychloride	
1414	Acetylated distarch phosphate	
1413	Phosphated distarch phosphate	
1420	Starch acetate esterified with acetic anhydride	
1421	Starch acetate esterified with vinyl acetate	
1422	Acetylated distarch adipate	
1440	Hydroxypropyl starch	
1442	Hydroxypropyl starch phosphate	

5.   HYGIENE AND HANDLING

5.1   The final product shall be free from any foreign material that poses a threat to human health.

5.2   When tested by appropriate methods of sampling and examination prescribed by the Codex Alimentarius Commission, the product:

   (i)   shall be free from microorganisms or substances originating from microorganisms in amounts which may present a hazard to health in accordance with standards established by the Codex Alimentarius Commission;

   (ii)   shall not contain histamine that exceeds 20 mg/100 g. This applies only to species of *Clupeidae, Scombridae, Scombresocidae, Pomatomidae* and *Coryphaenedae* families;

(iii)   shall not contain any other substance in amounts which may present a hazard to health in accordance with standards established by the Codex Alimentarius Commission.

5.3   It is recommended that the products covered by the provisions of this standard be prepared and handled in accordance with the appropriate sections of the Recommended International Code of Practice - General Principles of Food Hygiene (CAC/RCP 1-1969, Rev. 3-1997) and the following relevant Codes:

(i)   the Recommended International Code of Practice for Frozen Fish (CAC/RCP 16-1978);

(ii)   the Recommended International Code of Practice for Frozen Battered and/or Breaded Fishery Products (CAC/RCP 35-1985);

(iii)   the Recommended International Code of Practice for Minced Fish Prepared by Mechanical Separation (CAC/RCP 27-1983).

(iv)   The Recommended International Code of Practice for the Processing and Handling of Quick Frozen Foods (CAC/RCP 8-1976).

6.   LABELLING

In addition to Sections 2, 3, 7 and 8 of the Codex General Standard for the Labelling of Prepackaged Foods (CODEX STAN 1-1985, Rev. 1-1991) the following specific provisions apply:

6.1   THE NAME OF THE FOOD

6.1.1   The name of the food to be declared on the label shall be "breaded" and/or "battered", "fish sticks" (fish fingers), "fish portions", or "fillets" as appropriate or other specific names used in accordance with the law and custom of the country in which the food is sold and in a manner so as not to confuse or mislead the consumer.

6.1.2   The label shall include reference to the species or mixture of species.

6.1.3   In addition there shall appear on the label either the term "quick frozen" or the term "frozen" whichever is customarily used in the country in which the food is sold, to describe a product subjected to the freezing processes as defined in subsection 2.2.

6.1.4   The label shall show whether the products are prepared from minced fish flesh, fish fillets or a mixture of both in accordance with the law and custom of the country in which the food is sold and in a manner so as not to confuse or mislead the consumer.

6.1.5   The label shall state that the product should be maintained under conditions that will maintain the quality during transportation, storage and distribution.

6.2   STORAGE INSTRUCTIONS

The label shall include terms to indicate that the product shall be stored at a temperature of -18°C or colder.

6.3   LABELLING OF NON-RETAIL CONTAINERS

Information specified above shall be given either on the container or in accompanying documents, except that the name of the food, lot identification, and the name and address of the manufacturers or packer, as well as storage instructions, shall always appear on the container. However, lot identification, and the name and address may be replaced by an identification mark, provided that such a mark is clearly identifiable with the accompanying documents.

7. SAMPLING, EXAMINATION AND ANALYSIS

7.1 SAMPLING

(i) Sampling of lots for examination of the product shall be in accordance with the FAO/WHO Codex Alimentarius Sampling Plans for Prepackaged Foods (AQL-6.5) (CODEX STAN 233-1969). For prepackaged goods the sample unit is the entire container. For products packed in bulk the sample unit is at least 1 kg of fish sticks (fish finger), fish portions or fillets.

(ii) Sampling of lots for examination of net weight shall be carried out in accordance with an appropriate sampling plan meeting the criteria established by the Codex Alimentarius Commission.

7.2 DETERMINATION OF NET WEIGHT

The net weight (exclusive of packaging material) is determined on each whole primary container of each sample representing a lot and shall be determined in the frozen state.

7.3 SENSORY AND PHYSICAL EXAMINATION

Samples taken for sensory and physical examination shall be assessed by persons trained in such examination and in accordance with procedures elaborated in Sections 7.4 through 7.7, Annex A and the *Guidelines for the Sensory Evaluation of Fish and Shellfish in Laboratories (CAC/GL 31 - 1999)*.

7.4 ESTIMATION OF FISH CORE

According to A.O.A.C. Method 996.15.

7.5 DETERMINATION OF GELATINOUS CONDITIONS

According to the AOAC Methods - "Moisture in Meat and Meat Products, Preparation of Sample Procedure"; 983.18 and "Moisture in Meat" (Method A); 950.46.

7.6 ESTIMATION OF PROPORTION OF FISH FILLETS AND MINCED FISH FLESH

See Annex B.

7.7 COOKING METHODS

The frozen sample shall be cooked prior to sensory assessment according to the cooking instructions on the package. When such instructions are not given, or equipment to cook the sample according to the instructions is not obtainable, the frozen sample shall be cooked according to the applicable method(s) given below:

Use procedure 976.16 of the A.O.A.C. It is based on heating product to an internal temperature of 65-70°C. Cooking times vary according to size of product and equipment used. If determining cooking time, cook extra samples, using a temperature measuring device to determine internal temperature.

7.8 DETERMINATION OF HISTAMINE

According to the AOAC Methods 977.13.

8. DEFINITION OF DEFECTIVES

The sample unit shall be considered defective when it exhibits any of the properties defined below:

8.1 FOREIGN MATTER (COOKED STATE)

The presence in the sample unit of any matter which has not been derived from fish (excluding packing material), does not pose a threat to human health, and is readily recognized without magnification or is present at a level determined by any method including magnification that indicates non-compliance with good manufacturing and sanitation practices.

8.2   BONES (COOKED STATE) (IN PACKS DESIGNATED BONELESS)

More than one bone per kg greater or equal to 10 mm in length, or greater or equal to 1 mm in diameter; a bone less than or equal to 5 mm in length, is not considered a defect if its diameter is not more than 2 mm. The foot of a bone (where it has been attached to the vertebra) shall be disregarded if its width is less than or equal to 2 mm, or if it can easily be stripped off with a fingernail.

8.3   ODOUR AND FLAVOUR (COOKED STATE)

A sample unit affected by persistent and distinct objectionable odour and flavours indicative of decomposition, or rancidity or of feed.

8.4   FLESH ABNORMALITIES

Objectionable textural characteristics such as gelatinous conditions of the fish core together with greater than 86% moisture found in any individual fillet or sample unit with pasty texture resulting from parasites affecting more than 5% of the sample unit by weight.

9.   LOT ACCEPTANCE

A lot shall be considered as meeting the requirements of this standard when:

(i)   the total number of defectives as classified according to Section 8 does not exceed the acceptance number (c) of the appropriate sampling plan in the Sampling Plans for Prepackaged Foods (AQL-6.5) (CODEX STAN 233-1969);

(ii)   the average percent fish flesh of all sample units is not less than 50% of the frozen weight;

(iii)   the average net weight of all sample units is not less than the declared weight, provided there is no unreasonable shortage in any container; and

(iv)   the Food Additives, Hygiene and Labelling requirements of Sections 4, 5 and 6 are met.

"ANNEX A": SENSORY AND PHYSICAL EXAMINATION

The sample used for sensory evaluation should not be the same as that used for other examinations.

1. Complete net weight determination, according to defined procedures in Section 7.2.

2. Complete fish core determination on one set of the sample units according to defined procedures in Section 7.4.

3. Complete the estimation of the proportion of fillets and minced flesh, if required.

4. Cook the other set of sample units and examine for odour, flavour, texture, foreign matter, and bones.

5. In cases where a final decision on gelatinous conditions cannot be made in the thawed uncooked state, the disputed material is sectioned from the product and gelatinous condition confirmed by cooking as defined in Section 7.7 or by using the procedure in Section 7.5 to determine if greater than 86% moisture is present in any product unit. If a cooking evaluation is inconclusive, then procedure in 7.5 would be used to make the exact determination of moisture content.

ANNEX B: ESTIMATION OF PROPORTION OF FISH FILLETS AND MINCED FISH FLESH

(West European Fish Technologists Association - WEFTA Method)

a)   Equipment

Balance, sensitive to 0.1 g

Circular sieve - 200 mm diameter, 2.5 or 2.8 mesh opening (ISO) soft rubber edge (or blunt) spatula, forks, suitable sized plates, water tight plastic bags.

b)   Preparation of Samples

Fish Portions/Sticks: Take as many portions as needed to provide a fish core sample of about 200g (2kg). If breaded and/or battered firrst strip coating according to the method describer in section 7.4.

c)   Detemination of Weights "A" of the Frozen Fish Samples

Weight the single fish portions/decoated fish cores while they are still frozen. Smaller portions are combined to a sample sub-units of about 200 g (e.g. 10). fish sticks of about 20 g each). Record the weight "A" n of the sub-units. Place the pre-weighed sample sub-units into water tight bags.

d)   Thawing

Thaw the samples by immersing the bags into a gently agitated water bath of about 20°C, but not more than 35°C.

e)   Draining

After thawing has been completed (duration about 20-30 min.) take each sample unit, one at a time, and drain the exuded fluid (thaw drip) for 2 minutes on a pre-weighed circular sieve incluned at an angle of 17-20 degrees. Remove adhering drip from the bottom of the sieve by use of a paper towel when draining is completed.

f)   Determination of weight "B" of the Drained Fish Sample and Weight "C" of the Thaw Drip

Determine the weight of the drained fish sample "B" - sieve plus fish minus sieve weight. The difference of "A" - "B" is the weight of exuded fluid - thaw drip.

g)   Separation

Place the drained fish core on a plate and separate the minced flesh from the fillet using a fork to hold the fillet flesh and a soft, rubber edge spatula to scrape off the minced flesh.

CODEX STANDARD FOR QUICK FROZEN SHRIMPS OR PRAWNS

CODEX STAN 92-1981, REV. 1- 1995

1. SCOPE

This standard applies to quick frozen raw or partially or fully cooked shrimps or prawns,[1] peeled or unpeeled.

2. DESCRIPTION

2.1 PRODUCT DEFINITION

2.1.1 Quick frozen shrimp is the product obtained from species of the following families:

    (a)   *Penaeidae*
    (b)   *Pandalidae*
    (c)   *Crangonidae*
    (d)   *Palaemonidae*

2.1.2   The pack shall not contain a mixture of genera but may contain a mixture of species of the same genus which have similar sensory properties.

2.2 PROCESS DEFINITION

The water used for cooking and cooling shall be of potable quality or clean seawater.

The product, after any suitable preparation, shall be subjected to a freezing process and shall comply with the conditions laid down hereafter. The freezing process shall be carried out in appropriate equipment in such a way that the range of temperature of maximum crystallization is passed quickly. The quick freezing process shall not be regarded as complete unless and until the product temperature has reached -18°C or colder at the thermal centre after thermal stabilization. The product shall be kept deep frozen so as to maintain the quality during transportation, storage and distribution.

Quick frozen shrimps shall be processed and packaged so as to minimize dehydration and oxidation.

2.3 PRESENTATION

2.3.1   Any presentation of the product shall be permitted provided that it:

2.3.1.1 meets all requirements of this standard; and

2.3.1.2 is adequately described on the label to avoid confusing or misleading the consumer.

2.3.2   The shrimp may be packed by count per unit of weight or per package.

3. ESSENTIAL COMPOSITION AND QUALITY FACTORS

3.1 SHRIMP

Quick frozen shrimp shall be prepared from sound shrimp which are of a quality fit to be sold fresh for human consumption.

3.2 GLAZING

If glazed, the water used for glazing or preparing glazing solutions shall be of potable quality or shall be clean sea-water. Potable water is fresh-water fit for human consumption. Standards of potability

---

[1]   Hereafter referred to as shrimp.

shall not be less than those contained in the latest edition of the WHO "International Guidelines for Drinking Water Quality". Clean sea-water is sea-water which meets the same microbiological standards as potable water and is free from objectionable substances.

3.3 OTHER INGREDIENTS

All other ingredients used shall be of food grade quality and conform to all applicable Codex standards.

3.4 FINAL PRODUCT

Products shall meet the requirements of this standard when lots examined in accordance with Section 9 comply with the provisions set out in Section 8. Products shall be examined by the methods given in Section 7.

4. FOOD ADDITIVES

Only the use of the following additives is permitted.

Additive	Maximum Level in the final product
Acidity Regulators	
330 Citric acid	GMP
450(iii) Tetrasodium diphosphate	10 g/kg expressed as
450(v) Tetrapotassium diphosphate	P_2O_5, singly or in
451(i) Pentasodium triphosphate	combination (includes
451(ii) Pentapotassium triphosphate	natural phosphate)
Antioxidant	
300 Ascorbic acid (L-)	GMP
Colours	
124 Ponceau 4R	30 mg/kg in heat-treated products only
Preservatives	
221 Sodium sulphite	100 mg/kg in the edible
223 Sodium metabisulphite	part of the raw product,
224 Potassium metabisulphite	or 30 mg/kg in the
225 Potassium sulphite	edible part of the cooked product, singly or in combination, expressed as SO_2

5. HYGIENE AND HANDLING

5.1 The final product shall be free from any foreign material that poses a threat to human health.

5.2 When tested by appropriate methods of sampling and examination prescribed by the Codex Alimentarius Commission , the product:

    (i)   shall be free from microorganisms or substances originating from microorganisms in amounts which may present a hazard to health in accordance with standards established by the Codex Alimentarius Commission;

(ii)   shall not contain any other substance in amounts which may present a hazard to health in accordance with standards established by the Codex Alimentarius Commission.

5.3   It is recommended that the products covered by the provisions of this standard be prepared and handled in accordance with the appropriate sections of the Recommended International Code of Practice - General Principles of Food Hygiene (CAC/RCP 1-1969, Rev. 3-1997) and the following relevant Codes:

(i)   the Recommended International Code of Practice for Frozen Fish (CAC/RCP 16-1978);

(ii)   the Recommended International Code of Practice for Frozen Shrimps or Prawns (CAC/RCP 17-1978 and Supplement November 1989);

(iii)   the Recommended International Code of Practice for the Processing and Handling of Quick Frozen Foods (CAC/RCP 8-1976).

(iv)   The sections on the Products of Aquaculture in the Proposed Draft International Code of Practice for Fish and Fishery Products (under elaboration)[2]

6.   LABELLING

In addition to the provisions of the Codex General Standard for the Labelling of Prepackaged Foods (CODEX STAN 1-1985, Rev.1 - 1991) the following specific provisions apply:

6.1   THE NAME OF THE FOOD

The name of the product as declared on the label shall be "shrimps" or "prawns" according to the law, custom or practice in the country in which the product is to be distributed.

6.1.1   There shall appear on the label, reference to the presentation in close proximity to the name of the product in such descriptive terms that will adequately and fully describe the nature of the presentation of the product to avoid misleading or confusing the consumer.

6.1.2   In addition to the specified labelling designations above, the usual or common trade names of the variety may be added so long as it is not misleading to the consumer in the country in which the product will be distributed.

6.1.3   Products shall be designated as cooked, or partially cooked, or raw as appropriate.

6.1.4   If the product has been glazed with sea-water, a statement to this effect shall be made.

6.1.5   The term "quick frozen", shall also appear on the label, except that the term "frozen" may be applied in countries where this term is customarily used for describing the product processed in accordance with subsection 2.2 of this standard.

6.1.6   The label shall state that the product should be maintained under conditions that will maintain the quality during transportation, storage and distribution.

6.2   NET CONTENTS (GLAZED PRODUCTS)

Where the food has been glazed the declaration of net contents of the food shall be exclusive of the glaze.

6.3   STORAGE INSTRUCTIONS

The label shall include terms to indicate that the product shall be stored at a temperature of -18°C or colder.

---

[2]   The Proposed Draft Code of Practice, when finalized, will replace all current Codes of Practice for Fish and Fishery Products

6.4   LABELLING OF NON-RETAIL CONTAINERS

Information specified above shall be given either on the container or in accompanying documents, except that the name of the food, lot identification, and the name and address as well as storage instructions shall always appear on the container.

However, lot identification, and the name and address may be replaced by an identification mark, provided that such a mark is clearly identifiable with the accompanying documents.

7.   SAMPLING, EXAMINATION AND ANALYSES

7.1   SAMPLING

(i)   Sampling of lots for examination of the product shall be in accordance with the FAO/WHO Codex Alimentarius Sampling Plans for Prepackaged Foods (AQL - 6.5) (CODEX STAN 233-1969). The sample unit is the primary container or for individually quick frozen products is at least a 1 kg portion of the sample unit.

(ii)   Sampling of lots for examination of net weight shall be carried out in accordance with an appropriate sampling plan meeting the criteria established by the Codex Alimentarius Commission.

7.2   SENSORY AND PHYSICAL EXAMINATION

Samples taken for sensory and physical examination shall be assessed by persons trained in such examination and in accordance with procedures elaborated in Sections 7.3 through 7.6, Annex A and the Guidelines for the Sensory Evaluation of Fish and Shellfish in Laboratories (CAC/GL 31 - 1999).

7.3   DETERMINATION OF NET WEIGHT

7.3.1   Determination of net weight of Products not Covered by Glaze

The net weight (exclusive of packaging material) of each sample unit representing a lot shall be determined in the frozen state.

7.3.2   Determination of Net Weight of Products Covered by Glaze

Procedure

(1)   Open the package with quick frozen shrimps or prawns imediately after removal from low temperature storage.

(i)   For the raw product, place the contents in a container into which fresh water at room temperature is introduced from the bottom at a flow of approximately 25 litres per minute.

(ii)   For the cooked product place the product in a container containing an amount of fresh potable water of 27°C (80° F) equal to 8 times the declared weight of the product. Leave the product in the water until all ice is melted. If the product is block frozen, turn block over several times during thawing. The point at which thawing is complete can be determined by gently probing the block apart.

(2)   Weigh a dry clean sieve with woven wire cloth with nominal size of the square aperture 2.8 mm (ISO Recommendation R565) or alternatively 2.38 mm (US No. 8 Standard Screen).

(i)   If the quantity of the total contents of the package is 500 g (1.1 lbs) or less, use a sieve with a diameter of 20 cm (8 inches).

(ii)   If the quantity of the total contents of the package is more than 500 g (1.1 lbs) use a sieve with a diameter of 30 cm (12 inches).

(3)   After all glaze that can be seen or felt has been removed and the shrimps or prawns separate easily, empty the contents of the container on the previously weighed sieve. Incline the sieve at an angle of about 20° and drain for two minutes

(4)   Weigh the sieve containing the drained product. Subtract the mass of the sieve; the resultant figure shall be considered to be the net content of the package.

7.4   DETERMINATION OF COUNT

When declared on the label, the count of shrimp shall be determined by counting the numbers of shrimp in the container or a representative sample thereof and dividing the count of shrimp by the actual de-glazed weight to determine the count per unit weight.

7.5   PROCEDURES FOR THAWING

The sample unit is thawed by enclosing it in a film type bag and immersing in water at room temperature (not greater than 35°C). The complete thawing of the product is determined by gently squeezing the bag occasionally so as not to damage the texture of the shrimp, until no hard core or ice crystals are left.

7.6   COOKING METHODS

The following procedures are based on heating the product to an internal temperature of 65-70°C. The product must not be overcooked. Cooking times vary according to the size of the product and the temperature used. The exact times and conditions of cooking for the product should be determined by prior experimentation.

Baking Procedure: Wrap the product in aluminum foil and place it evenly on a flat cookie sheet or shallow flat pan.

Steaming Procedure: Wrap the product in aluminum foil and place it on a wire rack suspended over boiling water in a covered container.

Boil-in-Bag Procedure: Place the product into a boilable film-type pouch and seal. Immerse the pouch into boiling water and cook.

Microwave Procedure: Enclose the product in a container suitable for microwave cooking. If plastic bags are used, check to ensure that no odour is imparted from the plastic bags. Cook according to equipment instructions.

8.   DEFINITION OF DEFECTIVES

The sample unit shall be considered as defective when it exhibits any of the properties defined below.

8.1   DEEP DEHYDRATION

Greater than 10% of the weight of the shrimp in the sample unit or greater than 10% of the surface area of the block exhibits excessive loss of moisture clearly shown as white or yellow abnormality on the surface which masks the colour of the flesh and penetrates below the surface, and cannot be easily removed by scraping with a knife or other sharp instrument without unduly affecting the appearance of the shrimp.

8.2   FOREIGN MATTER

The presence in the sample unit of any matter which has not been derived from shrimp does not pose a threat to human health, and is readily recognized without magnification or is present at a level determined by any method including magnification, that indicates non-compliance with good manufacturing and sanitation practices.

8.3   ODOUR/FLAVOUR

Shrimp affected by persistent and distinct objectionable odours or flavours indicative of decomposition or rancidity or of feed.

8.4   DISCOLOURATION

Distinct blackening or green or yellow discoloration, singly or in combination of more than 10% of the surface area of individual shrimp which affects more than 25% of the sample unit.

9.   LOT ACCEPTANCE

A lot shall be considered as meeting the requirements of this standard when:

(i)   the total number of defectives as classified according to section 8 does not exceed the acceptance number (c) of the appropriate sampling plan in the Sampling Plans for Prepackaged Foods (AQL-6.5) (CODEX STAN 233-1969);

(ii)   the total number of sample units not meeting the count designation as defined in section 2.3 does not exceed the acceptance number (c) of the appropriate sampling plan in the Sampling Plans for Prepackaged Foods (AQL - 6.5) (CODEX STAN 233-1969).;

(iii)   the average net weight of all sample units is not less than the declared weight, provided there is no unreasonable shortage in any individual container;

(iv)   the Food Additives, Hygiene and Labelling requirements of Sections 4, 5 and 6 are met.

"ANNEX A": SENSORY AND PHYSICAL EXAMINATION

1. Complete net weight determination, according to defined procedures in Section 7.3 (de-glaze as required).

2. Examine the frozen shrimp in the sample unit or the surface of the block for the presence of dehydration. Determine the percentage of shrimp or surface area affected.

3. Thaw using the procedure described in Section 7.5 and individually examine each shrimp in the sample unit for the presence of foreign matter and presentation defects. Determine the weight of shrimp affected by presentation defects.

4. Examine product for count declarations in accordance with procedures in Section 7.4.

5. Assess the shrimp for odour and discolouration as required.

6. In cases where a final decision regarding the odour/flavour cannot be made in the thawed state, a small portion of the sample unit (100 to 200 g) is prepared without delay for cooking and the odour/flavour confirmed by using one of the cooking methods defined in Section 7.6.

CODEX STANDARD FOR QUICK FROZEN LOBSTERS

CODEX STAN 95 - 1981, REV 1 - 1995

1. SCOPE

This standard applies to quick frozen raw or cooked lobsters, rock lobsters, spiny lobsters and slipper lobsters.[1]

2. DESCRIPTION

2.1 PRODUCT DEFINITION

2.1.1 The product is prepared from lobster from the genus *Homarus* of the family *Nephropidae* and from the families *Palinuridae* and *Scyllaridae*. It may also be prepared from *Nephrops norvegicus* provided it is presented as Norway lobster.

2.1.2 The pack shall not contain a mixture of species.

2.2 PROCESS DEFINITION

The water used for cooking shall be of potable quality or clean seawater.

The product, after any suitable preparation, shall be subjected to a freezing process and shall comply with the conditions laid down hereafter. The freezing process shall be carried out in appropriate equipment in such a way that the range of temperature of maximum crystallization is passed quickly. The quick freezing process shall not be regarded as complete unless and until the product temperature has reached -18°C or colder at the thermal centre after thermal stabilization. The product shall be kept deep frozen so as to maintain the quality during transportation, storage and distribution.

Quick frozen lobsters shall be processed and packaged so as to minimize dehydration and oxidation.

2.3. PRESENTATION

2.3.1 Any presentation of the product shall be permitted provided that it:

2.3.1.1 meets all requirements of this standard;

2.3.1.2 is adequately described on the label to avoid confusing or misleading the consumer.

2.3.2 The lobster may be packed by count per unit of weight or per package or within a stated weight range.

3. ESSENTIAL COMPOSITION AND QUALITY FACTORS

3.1 LOBSTERS

The product shall be prepared from sound lobsters which are of a quality fit to be sold fresh for human consumption.

3.2 GLAZING

If glazed, the water used for glazing or preparing glazing solutions shall be of potable quality or

---

[1]   Hereafter referred to as lobster.

shall not be less than those contained in the latest edition of the WHO "International Guidelines for Drinking Water Quality". Clean sea-water is sea-water which meets the same microbiological standards as potable water and is free from objectionable substances.

3.3   OTHER INGREDIENTS

All other ingredients used shall be of food grade quality and conform to all applicable Codex standards.

3.4   FINAL PRODUCT

Products shall meet the requirements of this standard when lots examined in accordance with Section 9 comply with the provisions set out in Section 8. Products shall be examined by the methods given in Section 7.

4.   FOOD ADDITIVES

Only the use of the following additives is permitted.

Additive	Maximum Level in the Final Product
Moisture/Water Retention Agents	
451(i) Pentasodium triphosphate	10 g/kg expressed as P_2O_5, singly or in combination (includes natural phosphate)
451(ii) Pentapotassium triphosphate	
452(i) Sodium polyphosphate	
452(iv) Calcium polyphosphates	
Preservatives	
221 Sodium sulphite	100 mg/kg in the edible part of the raw product, or 30 mg/kg in the edible part of the cooked product, singly or in combination, expressed as SO_2
223 Sodium metabisulphite	
224 Potassium metabisulphite	
225 Potassium sulphite	
228 Potassium bisulphite (for use in the raw product only)	
Antioxidants	
300 Ascorbic acid	GMP
301 Sodium ascorbate	
303 Potassium ascorbate	

5.   HYGIENE AND HANDLING

5.1   The final product shall be free from any foreign material that poses a threat to human health.

5.2   When tested by appropriate methods of sampling and examination prescribed by the Codex Alimentarius Commission , the product:

(i)   shall be free from microorganisms or substances originating from microorganisms in amounts which may present a hazard to health in accordance with standards established by the Codex Alimentarius Commission;

(ii)   shall not contain any other substance in amounts which may present a hazard to health in accordance with standards established by the Codex Alimentarius Commission.

5.3   It is recommended that the products covered by the provisions of this standard be prepared and handled in accordance with the appropriate sections of the Recommended International Code of Practice - General Principles of Food Hygiene (CAC/RCP 1-1969, Rev. 3-1997) and the following relevant Codes:

(i)   The Recommended International Code of Practice for Lobsters (CAC/RCP 24-1978);

(ii)   The Recommended International Code of Practice for the Processing and Handling of Quick Frozen Foods (CAC/RCP 8-1976);

(iii)   The sections on the Products of Aquaculture in the Proposed Draft International Code of Practice for Fish and Fishery Products (under elaboration).[2]

6.   LABELLING

In addition to the provisions of the General Standard for the Labelling of Prepackaged Foods (CODEX STAN 1-1985, Rev. 1-1991) the following specific provisions apply:

6.1   THE NAME OF THE FOOD

The product shall be designated:

(i)   Lobster if derived from the genus *Homarus*;

(ii)   Rock Lobster, Spiny Lobster or Crawfish if derived from species of the family *Palinuridae*;

(iii)   Slipper Lobster, Bay Lobster or Sand Lobster if derived from species of the family *Scyllaridae*;

(iv)   Norway Lobster if derived from the species *Nephrops norvegicus*.

6.1.1   There shall appear on the label, reference to the form of presentation in close proximity to the name of the product in such descriptive terms that will adequately and fully describe the nature of the presentation of the product to avoid misleading or confusing the consumer.

6.1.2   In addition to the specified labelling designations above, the usual or common trade names of the variety may be added so long as it is not misleading to the consumer in the country in which the product will be distributed.

6.1.3   Products shall be designated as cooked or raw as appropriate.

6.1.4   If the product has been glazed with sea-water, a statement to this effect shall be made.

6.1.5   The term "quick frozen", shall also appear on the label, except that the term "frozen" may be applied in countries where this term is customarily used for describing the product processed in accordance with subsection 2.2 of this standard.

6.1.6   The label shall state that the product should be maintained under conditions that will maintain the quality during transportation, storage and distribution.

---

[2]   The Proposed Draft Code of Practice, when finalized, will replace all current Codes of Practice for Fish and Fishery Products

6.2   NET CONTENTS (GLAZED PRODUCTS)

Where the food has been glazed the declaration of net contents of the food shall be exclusive of the glaze.

6.3   STORAGE INSTRUCTIONS

The label shall include terms to indicate that the product shall be stored at a temperature of -18°C or colder.

6.4   LABELLING OF NON-RETAIL CONTAINERS

Information specified above shall be given either on the container or in accompanying documents, except that the name of the food, lot identification, and the name and address of the manufacturer or packer as well as storage instructions shall always appear on the container.

However, lot identification, and the name and address may be replaced by an identification mark, provided that such a mark is clearly identifiable with the accompanying documents.

7.   SAMPLING, EXAMINATION AND ANALYSES

7.1   SAMPLING

(i)   Sampling of lots for examination of the product shall be in accordance with the FAO/WHO Codex Alimentarius Sampling Plans for Prepackaged Foods (AQL - 6.5) (CODEX STAN 233-1969). In the case of shell on lobster the sample unit is an individual lobster. In the case of shell-off lobster the sample unit shall be at least a 1 kg portion of lobster from the primary container.

(ii)   Sampling of lots for examination of net weight shall be carried out in accordance with an appropriate sampling plan meeting the criteria established by the Codex Alimentarius Commission.

7.2   SENSORY AND PHYSICAL EXAMINATION

Samples taken for sensory and physical examination shall be assessed by persons trained in such examination and using procedures elaborated in Sections 7.3 through 7.6, Annex A and the *Guidelines for the Sensory Evaluation of Fish and Shellfish in Laboratories (CAC/GL 31 - 1999)*.

7.3   DETERMINATION OF NET WEIGHT

7.3.1   Determination of net weight of Products not Covered by Glaze

The net weight (exclusive of packaging material) of each sample unit representing a lot shall be determined in the frozen state.

7.3.2   Determination of Net Weight of Products Covered by Glaze

(Alternate Methods)

(1)   As soon as the package is removed from frozen temperature storage, open immediately and place the contents under a gentle spray of cold water until all ice glaze that can be seen or felt is removed. Remove adhering water by the use of paper towel and weigh the product.

(2)   The pre-weighed glazed sample is immersed into a water bath by hand, until all glaze is removed, which preferably can be felt by the fingers. As soon as the surface becomes rough, the still frozen sample is removed from the water bath and dried by use of a paper towel before estimating the net product content by second weighing. By this procedure thaw drip losses and/or re-freezing of adhering moisture can be avoided.

(3)   (i)   As soon as the package is removed from frozen temperature storage, place the product in a container containing an amount of fresh potable water of 27°C (80°F) equal to 8 times the declared weight of the product. Leave the product in the water until all ice is melted. If the product is block frozen, turn block over several time during thawing. The point at which thawing is complete can be determined by gently probing the block.

(ii)   Weigh a dry clean sieve with woven wire cloth with nominal size of the square aperture 2.8 mm (ISO Recommendation R565) or alternatively 2.38 mm (U.S. No. 8 Standard Screen.)

   (a)   If the quantity of the total contents of the package is 500 g (1.1 lbs) or less, use a sieve with a diameter of 20 cm (8 inches).

   (b)   If the quantity of the total contents of the package is more than 500 g (1.1 lbs) use a sieve with a diameter of 30 cm (12 inches).

(iii)   After all glaze that can be seen or felt has been removed and the lobsters separate easily, empty the contents of the container on the previously weighed sieve. Incline the sieve at an angle of about 20° and drain for two minutes.

(iv)   Weigh the sieve containing the drained product. Subtract the mass of the sieve; the resultant figure shall be considered to be part of the net content of the package.

7.4   DETERMINATION OF COUNT

When declared on the label, the count shall be determined by counting all lobsters or tails in the primary container and dividing the count of lobster by the average deglazed weight to determine the count per unit weight.

7.5   PROCEDURE FOR THAWING

The sample unit is thawed by enclosing it in a film type bag and immersing in water at room temperature (not greater than 35°C). The complete thawing of the product is determined by gently squeezing the bag occasionally so as not to damage the texture of the lobster, until no hard core or ice crystals are left.

7.6   COOKING METHODS

The following procedures are based on heating the product to an internal temperature of 65-70°C. The product must not be overcooked. Cooking times vary according to the size of the product and the temperature used. The exact times and conditions of cooking for the product should be determined by prior experimentation.

Baking Procedure: Wrap the product in aluminum foil and place it evenly on a flat cookie sheet or shallow flat pan.

Steaming Procedure: Wrap the product in aluminum foil and place it on a wire rack suspended over boiling water in a covered container.

Boil-in-Bag Procedure: Place the product into a boilable film-type pouch and seal. Immerse the pouch into boiling water and cook.

Microwave Procedure: Enclose the product in a container suitable for microwave cooking. If plastic bags are used check to ensure that no odour is imparted from the plastic bags. Cook according to equipment specifications.

8. DEFINITION OF DEFECTIVES

The sample unit shall be considered as defective when it exhibits any of the properties defined below.

8.1 DEEP DEHYDRATION

Greater than 10% of the weight of the lobster in the sample unit or greater than 10% of the surface area of the block exhibits excessive loss of moisture clearly shown as white or yellow abnormality on the surface which masks the colour of the flesh and penetrates below the surface, and cannot be easily removed by scraping with a knife or other sharp instrument without unduly affecting the appearance of the lobster.

8.2 FOREIGN MATTER

The presence in the sample unit of any matter which has not been derived from lobster, does not pose a threat to human health, and is readily recognized without magnification or is present at a level determined by any method including magnification that indicates non-compliance with good manufacturing and sanitation practices.

8.3 ODOUR/FLAVOUR

Lobster affected by persistent and distinct objectionable odours or flavours indicative of decomposition or rancidity, or feed.

8.4 DISCOLOURATION

Distinct blackening of more than 10% of the surface area of the shell of individual whole or half lobster, or in the case of tail meat and meat presentations distinct black, brown, green or yellow discolourations singly or in combination, of the meat affecting more than 10% of the declared weight.

9. LOT ACCEPTANCE

A lot shall be considered as meeting the requirements of this standard when:

(i)   the total number of defectives as classified according to section 8 does not exceed the acceptance number (c) of the appropriate sampling plan in the Sampling Plans for Prepackaged Foods (AQL-6.5) (CODEX STAN 233-1969);

(ii)   the total number of sample units not meeting the count or weight range designation as defined in Section 2.3 does not exceed the acceptance number (c) of the appropriate sampling plan in the Sampling Plans for Prepackaged Foods (AQL - 6.5) (CODEX STAN 233-1969);

(iii)   the average net weight of all sample units is not less than the declared weight, provided there is no unreasonable shortage in any individual container;

(iv)   the Food Additives, Hygiene and Labelling requirements of Sections 4, 5 and 6 are met.

"ANNEX A": SENSORY AND PHYSICAL EXAMINATION

1. Complete net weight determination, according to defined procedures in Section 7.3 (de-glaze as required).

2. Examine the frozen lobster for the presence of deep dehydration. Determine the percentage of lobster affected.

3. Thaw using the procedure described in Section 7.5 and individually examine each sample unit for the presence of foreign and objectionable matter.

4. Examine product count and weight declarations in accordance with procedures in Section 7.4.

5. Assess the lobster for odour and discolouration as required.

6. In cases where a final decision regarding the odour/flavour cannot be made in the thawed state, a small portion of the sample unit (100 to 200 g) is prepared without delay for cooking and the odour/flavour confirmed by using one of the cooking methods defined in Section 7.6.

CODEX STANDARD FOR QUICK FROZEN RAW SQUID

CODEX STAN 191 - 1995

1.   SCOPE

This standard applies to quick frozen raw squid and parts of raw squid, as defined below and offered for direct consumption without further processing. It does not apply to products indicated as intended for further processing or for other industrial purpose.

2.   DESCRIPTION

2.1   PRODUCT DEFINITION

Quick frozen squid and parts of squid are obtained from squid species of the following families:

(i)   *Loliginidae*

(ii)   *Ommastrephidae*

2.2   PROCESS DEFINITION

The product after any suitable preparation shall be subjected to a freezing process and shall comply with the conditions laid down hereafter. The freezing process shall be carried out in appropriate equipment in such a way that the range of temperature of maximum crystallization is passed quickly. The quick freezing process shall not be regarded as complete unless and until the product temperature has reached -18°C or colder at the thermal centre after thermal stabilization. The product shall be kept deep frozen so as to maintain the quality during transportation, storage and distribution.

Industrial repacking of intermediate quick frozen material under controlled conditions which maintain the quality of the product, followed by the reapplication of the quick freezing process as defined above is permitted.

Quick frozen squid and parts of squid shall be processed and packaged so as to minimize dehydration and oxidation.

2.3   PRESENTATION

Any presentation of the product shall be permitted provided that it:

(i)   meets all the requirements of this standard, and

(ii)   is adequately described on the label to avoid confusing or misleading the consumer.

3.   ESSENTIAL COMPOSITION AND QUALITY FACTORS

3.1   SQUID

Quick frozen squid shall be prepared from sound squid which are of a quality fit to be sold fresh for human consumption.

3.2   GLAZING

If glazed, the water used for glazing or preparing glazing solutions shall be of potable quality or shall be clean sea-water. Potable water is fresh-water fit for human consumption. Standards of potability shall not be less than those contained in the latest edition of the WHO "International Guidelines for Drinking Water Quality". Clean sea-water is sea-water which meets the same microbiological standards as potable water and is free from objectionable substances.

3.3   FINAL PRODUCT

Products shall meet the reguirements of this standard when lots examined in accordance with Section 9 comply with the provisions set out in Section 8. Products shall be examined by the methods given in Section 7.

4.   FOOD ADDITIVES

No food additives are permitted in these products.

5.   HYGIENE AND HANDLING

5.1   The final product shall be free from any foreign material that poses a threat to human health.

5.2   When tested by appropriate methods of sampling and examination prescribed by the Codex Alimentarius Commission , the product:

(i)   shall be free from microorganisms or substances originating from microorganisms in amounts which may present a hazard to health in accordance with standards established by the Codex Alimentarius Commission; and

(ii)   shall not contain any other substance in amounts which may present a hazard to health in accordance with standards established by the Codex Alimentarius Commission.

5.3   It is recommended that the product covered by the provisions of this standard be prepared and handled in accordance with the appropriate sections of the Recommended International Code of Practice - General Principles of Food Hygiene (CAC/RCP 1-1969, Rev. 3-1997) and the following relevant Codes:

(i)   the Recommended International Code of Practice for Frozen Fish (CAC/RCP 16-1978);

(ii)   the Recommended International Code of Practice for the Processing and Handling of Quick Frozen Foods (CAC/RCP 8-1976);

(iii)   the Recommended International Code of Practice for Cephalopods (CAC/RCP 37-1989).

6.   LABELLING

In addition to the provisions of the Codex General Standard for the Labelling of Prepackaged Foods (CODEX STAN 1-1985, Rev. 1-1991) the following specific provisions apply:

6.1   THE NAME OF THE FOOD

6.1.1   The name of the product shall be "squid", or another name according to the law, custom or practice in the country in which the product is to be distributed.

6.1.2   There shall appear on the label reference to the presentation, in close proximity to the name of the food in such additional words or phrases that will avoid misleading or confusing the consumer.

6.1.3   In addition, the labelling shall show the term "frozen", or "quick frozen" whichever is customarily used in the country in which the product is distributed, to describe a product subjected to the freezing process described in sub-section 2.2.

6.1.4   The label shall state that the product should be maintained under conditions that will maintain the quality during transportation, storage and distribution.

6.1.5   If the product has been glazed with sea-water, a statement to this effect shall be made.

6.2   NET CONTENTS (GLAZED PRODUCTS)

Where the food has been glazed, the declaration of net contents of the food shall be exclusive of the glaze.

6.3 STORAGE INSTRUCTIONS

The label shall include terms to indicate that the product shall be stored at a temperature of -18°C or colder.

6.4 LABELLING OF NON-RETAIL CONTAINERS

Information specified above shall be given either on the container or in accompanying documents, except that the name of the food, lot identification, and the name and address of the manufacturer or packer as well as storage instructions shall always appear on the container.

However, lot identification, and the name and address may be replaced by an identification mark, provided that such a mark is clearly identifiable with the accompanying documents.

7. SAMPLING, EXAMINATION AND ANALYSES

7.1 SAMPLING

7.1.1 Sampling of lots for examination of the product shall be in accordance with the FAO/WHO Codex Alimentarius Sampling Plans for Prepackaged Foods (AQL- 6.5) (CODEX STAN 233-1969). Sampling of lots composed of blocks shall be in accordance with the sampling plan developed for quick frozen fish blocks (reference to be provided). The sample unit is the primary container or for individually quick frozen products is at least 1 kg portion of the sample unit.

7.1.2 Sampling of lots for examination of net weight shall be carried out in accordance with an appropriate sampling plan meeting the criteria established by the Codex Alimentarius Commission.

7.2 SENSORY AND PHYSICAL EXAMINATION

Samples taken for sensory and physical examination shall be assessed by persons trained in such examination and in accordance with procedures elaborated in Sections 7.3 through 7.5, Annex A and the *Guidelines for the Sensory Evaluation of Fish and Shellfish in Laboratories (CAC/GL 31 - 1999)*.

7.3 DETERMINATION OF NET WEIGHT

7.3.1 Determination of Net Weight of Product not Covered by Glaze

The net weight (exclusive of packaging material) of each sample unit representing a lot shall be determined in the frozen state.

7.3.2 Determination of Net Weight of Products Covered by Glaze

(to be elaborated)

7.4 PROCEDURE FOR THAWING

The sample unit is thawed by enclosing it in a film-type bag and immersing in water at room temperature (not higher than 35°C). The complete thawing of the product is determined by gently squeezing the bag occasionally so as not to damage the texture of the squid until no hard core of ice crystals are left.

7.5 COOKING METHODS

The following procedures are based on heating the product to an internal temperature of 65-70°C. Cooking times vary according to the size of the product and the temperatures used. The exact times and conditions of cooking for the product should be determined by prior experimentation.

> *Baking Procedure:* Wrap the product in aluminum foil and place it evenly on a flat cookie sheet or shallow flat pan.

Steaming Procedure: Wrap the product in aluminum foil and place it on a wire rack suspended over boiling water in a covered container.

Boil-In-Bag Procedure: Place the product into a boilable film-type pouch and seal. Immerse the pouch into boiling water and cook.

Microwave Procedure: Enclose the product in a container suitable for microwave cooking. If plastic bags are used, check to ensure that no odour is imparted from the plastic bags. Cook according to equipment instructions.

8. DEFINITION OF DEFECTIVES

The sample unit shall be considered defective when it exhibit any of the properties defined below.

8.1 DEEP DEHYDRATION

Greater than 10% of the surface area of the sample unit exhibits excessive loss of moisture clearly shown as white or yellow abnormality on the surface which masks the colour of the flesh and penetrates below the surface, and cannot be easily removed by scraping with a knife or other sharp instrument without unduly affecting the appearance of the squid.

8.2 FOREIGN MATTER

The presence in the sample unit of any matter which has not been derived from squid (excluding packing material), does not pose a threat to human health, and is readily recognized without magnification or is present at a level determined by any method including magnification that indicates non-compliance with good manufacturing and sanitation practices.

8.3 ODOUR AND FLAVOUR

A sample unit affected by persistent and distinct objectional odours or flavours indicative of decomposition, which may be characterized also by light pinkish to red colour.

8.5 TEXTURE

Textural breakdown of the flesh, indicative of decomposition, characterized by muscle structure which is mushy or paste-like.

9. LOT ACCEPTANCE

A lot shall be considered as meeting the requirements of this standard when:

(i)   the total number of defectives as classified according to Section 8 does not exceed the acceptance number (c) of the appropriate sampling plan in the Sampling plans for Prepackaged Foods (AQL-6.5) (CODEX STAN 233-1969);

(ii)   the average net weight of all sample units is not less than the declared weight, provided there is no unreasonable shortage in any container;

(iii)   the Food Additives, Hygiene and Labelling requirements of Sections 4, 5 and 6 are met.

"ANNEX A": SENSORY AND PHYSICAL EXAMINATION

1.   Complete net weight determination, according to defined procedures in Section 7.3 (de-glaze as required).

2.   Examine the frozen squid for the presence of deep dehydration by measuring those areas which can only be removed with a knife or other sharp instrument. Measure the total surface area of the sample unit, and determine the percentage affected using the following formula;

$$\frac{\text{area affected}}{\text{total surface area}} \times 100\% = \% \text{ affected by deep dehydration}$$

3.   Thaw and individually examine each squid in the sample unit for the presence of foreign matter and colour.

4.   Examine each squid using the criteria outlined in Section 8. Flesh odours are examined by making a cut parallel to the surface of the flesh so that the exposed surface can be evaluated.

5.   In cases where a final decision on odour and texture can not be made in the thawed uncooked state, a portion of the sample unit is sectioned off and the odour, flavour and texture confirmed without delay by using one of the cooking methods defined in Section 7.5.

SECTION 2

CANNED FISH AND FISHERY PRODUCTS

CODEX STANDARD FOR CANNED FINFISH

CODEX STAN 119 - 1981, REV. 1 - 1995

1.   SCOPE

This standard applies to canned finfish packed in water, oil or other suitable packing medium. It does not apply to speciality products where the canned finfish constitutes less than 50% m/m of the net contents of the can or to canned finfish covered by other Codex product standards.

2.   DESCRIPTION

2.1   PRODUCT DEFINITION

Canned finfish is the product produced from the flesh of any species of finfish (other than canned finfish covered by other Codex product standards) which is suitable for human consumption and may contain a mixture of species, with similar sensoric properties, from within the same genus.

2.2   PROCESS DEFINITION

Canned finfish are packed in hermetically sealed containers and shall have received a processing treatment sufficient to ensure commercial sterility.

2.3   PRESENTATION

Any presentation of the product shall be permitted provided that it:

(i)   meets all requirements of this standard; and

(ii)   is adequately described on the label to avoid confusing or misleading the consumer.

3.   ESSENTIAL COMPOSITION AND QUALITY FACTORS

3.1   FISH

The product shall be prepared from sound finfish from which the heads, tails and viscera have been removed. The raw material shall be of a quality fit to be sold fresh for human consumption.

3.2   OTHER INGREDIENTS

The packing medium and all other ingredients used shall be of food grade quality and conform to all applicable Codex standards.

3.3.   DECOMPOSITION

Canned finfish of the families *Scombridae, Scombresocidae, Clupeidae, Coryphaenidae* and *Pomatomidae* shall not contain more than 10 mg/100 g of histamine based on the average of the sample units tested.

3.4   FINAL PRODUCT

Products shall meet the requirements of this Standard when lots examined in accordance with Section 9 comply with the provisions set out in Section 8. Products shall be examined by the methods given in Section 7.

4.   FOOD ADDITIVES

Additive	**Maximum Level in the Final Product**
Thickening or Gelling Agents (for use in packing media only)	

400	Alginic acid	GMP
401	Sodium alginate	
402	Potassium alginate	
404	Calcium alginate	
406	Agar	
407	Carrageenan and its Na, K, and NH_4 salts (including furcelleran)	
407a	Processed *Euchema* Seaweed (PES)	
410	Carob bean gum	
412	Guar gum	
413	Tragacanth gum	
415	Xanthan gum	
440	Pectins	
466	Sodium carboxymethylcellulose	

Modified Starches

1401	Acid treated starches	GMP
1402	Alkaline treated starches	
1404	Oxidized starches	
1410	Monostarch phosphate	
1412	Distarch phosphate esterified with sodium trimetaphosphate; esterified with phosphorus oxychloride	
1414	Acetylated distarch phosphate	
1413	Phosphated distarch phosphate	
1420	Starch acetate esterified with acetic anhydride	
1421	Starch acetate esterified with vinyl acetate	
1422	Acetylated distarch adipate	
1440	Hydroxypropyl starch	
1442	Hydroxypropyl starch phosphate	

Acidity Regulators

260	Acetic acid	GMP
270	Lactic acid (L-, D-, and DL-)	
330	Citric acid	

Natural Flavours

Spice oils	GMP
Spice extracts	
Smoke flavours (Natural smoke solutions and extracts)	

5. HYGIENE AND HANDLING

5.1   The final product shall be free from any foreign material that poses a threat to human health.

5.2   When tested by appropriate methods of sampling and examination prescribed by the Codex Alimentarius Commission, the product:

    (i)   shall be free from microorganisms capable of development under normal conditions of storage; and

    (ii)   no sample unit shall contain histamine that exceeds 20 mg per 100 g. This applies only to species of the families *Scombridae, Clupeidae, Coryphaenidae, Scombresocidae* and *Pomatomidae*.

    (iii)   shall not contain any other substance including substances derived from microorganisms in amounts which may represent a hazard to health in accordance with standards established by the Codex Alimentarius Commission; and

    (iv)   shall be free from container integrity defects which may compromise the hermetic seal.

5.3   It is recommended that the product covered by the provisions of this standard be prepared and handled in accordance with the appropriate sections of the Recommended International Code of Practice - General Principles of Food Hygiene (CAC/RCP 1-1969, Rev. 3-1997) and the following relevant Codes:

    (i)   the Recommended International Code of Practice for Canned Fish (CAC/RCP 10-1976);

    (ii)   the Recommended International Code of Hygienic Practice for Low-Acid and Acidified Low-Acid Canned Foods (CAC/RCP 23-1979, Rev. 2-1993);

    (iii)   The sections on the Products of Aquaculture in the Proposed Draft International Code of Practice for Fish and Fishery Products (under elaboration)[1]

6. LABELLING

In addition to the provisions of the Codex General Standard for the Labelling of Prepackaged Foods (CODEX STAN 1-1985, Rev. 1 - 1991) the following specific provisions apply.

6.1 NAME OF THE FOOD

6.1.1   The name of the product declared on the label shall be the common or usual name applied to the species in accordance with the law and custom of the country in which the product is sold, and in a manner not to mislead the consumer.

6.1.2   The name of the product shall be qualified by a term descriptive of the presentation.

6.1.3   The name of the packing medium shall form part of the name of the food.

6.1.4   Where a mixture of species of the same genus are used, they shall be indicated on the label.

6.1.5   In addition, the label shall include other descriptive terms that will avoid misleading or confusing the consumer.

---

[1]   The Proposed Draft Code of Practice, when finalized, will replace all current Codes of Practice for Fish and Fishery Products

7.   SAMPLING, EXAMINATION AND ANALYSES

7.1   SAMPLING

(i)   Sampling of lots for examination of the final product as prescribed in Section 3.3 shall be in accordance with the FAO/WHO Codex Alimentarius Sampling Plans for Prepackaged Foods (AQL-6.5) (Ref. CODEX STAN 233-1969).

(ii)   Sampling of lots for examination of net weight and drained weight, where appropriate, shall be carried out in accordance with an appropriate sampling plan meeting the criteria established by the Commission Alimentarius Commission.

7.2   SENSORIC AND PHYSICAL EXAMINATION

Samples taken for sensoric and physical examination shall be assessed by persons trained in such examination and in accordance with Sections 7.3 through 7.5, Annex A and the *Guidelines for the Sensory Evaluation of Fish and Shellfish in Laboratories (CAC/GL 31 - 1999)*.

7.3   DETERMINATION OF NET WEIGHT

The net weight of all sample units shall be determined by the following procedure:

(i)   Weigh the unopened container.

(ii)   Open the container and remove the contents.

(iii)   Weigh the empty container, (including the end) after removing excess liquid and adhering meat.

(iv)   Subtract the weight of the empty container from the weight of the unopened container. The resultant figure will be the net content.

7.4   DETERMINATION OF DRAINED WEIGHT

The drained weight of all sample units shall be determined by the following procedure:

(i)   Maintain the container at a temperature between 20°C and 30°C for a minimum of 12 hours prior to examination.

(ii)   Open and tilt the container to distribute the contents on a pre-weighed circular sieve which consists of wire mesh with square openings of 2.8 mm x 2.8 mm.

(iii)   Incline the sieve at an angle of approximately 17-20° and allow the fish to drain for two minutes, measured from the time the product is poured into the sieve.

(iv)   Weigh the sieve containing the drained fish.

(v)   The weight of drained fish is obtained by subtracting the weight of the sieve from the weight of the sieve and drained product.

7.5   DETERMINATION OF WASHED DRAINED WEIGHT (FOR PACKS WITH SAUCES)

(i)   Maintain the container at a temperature between 20°C and 30°C for a minimum of 12 hours prior to examination.

(ii)   Open and tilt the container and wash the covering sauce and then the full contents with hot tap water (approx. 40°C), using a wash bottle (e.g. plastic) on the tared circular sieve.

(iii)   Wash the contents of the sieve with hot water until free of adhering sauce; where necessary separate optional ingredients (spices, vegetables, fruits) with pincers. Incline the sieve at an angle of approximately 17-20° and allow the fish to drain two minutes, measured from the time the washing procedure has finished.

(iv)   Remove adhering water from the bottom of the sieve by use of paper towel. Weigh the sieve containing the washed drained fish.

(v)   The washed drained weight is obtained by subtracting the weight of the sieve from the weight of the sieve and drained product.

7.6.   DETERMINATION OF HISTAMINE

AOAC 977.13.

8.   DEFINITION OF DEFECTIVES

A sample unit will be considered defective when it exhibits any of the properties defined below.

8.1   FOREIGN MATTER

The presence in the sample unit of any matter, which has not been derived from fish or the packing medium, does not pose a threat to human health, and is readily recognized without magnification or is present at a level determined by any method including magnification that indicates non-compliance with good manufacturing and sanitation practices.

8.2   ODOUR/FLAVOUR

A sample unit affected by persistent and distinct objectionable odours or flavours indicative of decomposition or rancidity.

8.3   TEXTURE

(i)   Excessive mushy flesh uncharacteristic of the species in the presentation; or

(ii)   Excessively tough flesh uncharacteristic of the species in the presentation; or

(iii)   Honey combed flesh in excess of 5% of the drained contents.

8.4   DISCOLOURATION

A sample unit affected by distinct discolouration of the flesh indicative of decomposition or rancidity or by sulphide staining of more than 5% of the drained contents.

8.5   OBJECTIONABLE MATTER

A sample unit affected by Struvite crystals - any struvite crystal greater than 5 mm in length.

9.   LOT ACCEPTANCE

A lot shall be considered as meeting the requirements of this standard when:

(i)   the total number of defectives as classified according to Section 8 does not exceed the acceptance number (c) of the appropriate sampling plan in the Sampling Plans for Prepackaged Foods (AQL-6.5) (CODEX STAN 233-1969);

(ii)   the total number of sample units not meeting the presentation defined in 2.3 does not exceed the acceptance number (c) of the appropriate sampling plan in the Sampling Plans for Prepackaged Foods (AQL-6.5) (CODEX STAN 233-1969);

(iii)   the average net weight and the average drained weight where appropriate of all sample units examined is not less than the declared weight, and provided there is no unreasonable shortage in any individual container.

(iv)   the Food Additives, Hygiene and Handling and Labelling requirements of Sections 4, 5 and 6 are met.

ANNEX "A" :SENSORY AND PHYSICAL EXAMINATION

1. Complete external can examination for the presence of container integrity defects or can ends which may be distorted outwards.

2. Open can and complete weight determination according to defined procedures in Sections 7.3, 7.4 and 7.5.

3. Examine the product for the form of presentation.

4. Examine product for discolouration, foreign and objectionable matter. The presence of a hard bone is an indicator of underprocessing and will require an evaluation for sterility.

5. Assess odour, flavour and texture in accordance with the *Guidelines for the Sensory Evaluation of Fish and Shellfish in Laboratories (CAC/GL 31-1999)*

CODEX STANDARD FOR CANNED SALMON

CODEX STAN 3 - 1981, REV. 2 - 1995

1. SCOPE

This standard applies to canned salmon.

2. DESCRIPTION

2.1 PRODUCT DEFINITION

2.1.1 Canned Salmon is the product prepared from headed and eviscerated fish of any of the species listed below from which the fins and tails have been removed, and to which salt, water, salmon oil and/or other edible oils may have been added.

- *Salmo salar*

- *Oncorhynchus nerka*

- *Oncorhynchus kisutch*

- *Oncorhynchus tschawytscha*

- *Oncorhynchus gorbuscha*

- *Oncorhynchus keta*

- *Oncorhynchus masou*

2.2 PROCESS DEFINITION

Canned salmon is packed in hermetically sealed containers and shall have received a processing treatment sufficient to ensure commercial sterility.

2.3 PRESENTATION

2.3.1 Canned salmon shall consist of sections which are cut transversely from the fish and which are filled vertically into the can. The sections shall be packed so that the cut surfaces are approximately parallel with the ends of the container.

2.3.2 Any other presentation shall be permitted provided that it:

(i) is sufficiently distinctive from the form of presentation laid down under 2.3.1;

(ii) meets all other requirements of this standard; and

(iii) is adequately described on the label to avoid confusing or misleading the consumer.

3. ESSENTIAL COMPOSITION AND QUALITY FACTORS

3.1 SALMON

The product shall be prepared from sound fish of the species in Section 2.1 and of a quality fit to be sold fresh for human consumption.

3.2 OTHER INGREDIENTS

All other ingredients used shall be of food grade quality and conform to all applicable Codex standards.

3.3 FINAL PRODUCT

Products shall meet the requirements of this standard when lots examined in accordance with Section 9 comply with the provisions set out in Section 8. Products shall be examined by the methods given in Section 7.

4. FOOD ADDITIVES

No additives are permitted in this product.

5. HYGIENE AND HANDLING

5.1   The final product shall be free from any foreign material that poses a threat to human health.

5.2   When tested by appropriate methods of sampling and examination prescribed by the Codex Alimentarius Commission, the product:

(i)   shall be free from microorganisms capable of development under normal conditions of storage; and

(ii)   shall not contain any other substance derived from microorganisms in amounts which may represent a hazard to health in accordance with standards established by the Codex Alimentarius Commission; and

(iii)   shall be free from container integrity defects which may compromise the hermetic seal.

5.3   It is recommended that the product covered by the provisions of this standard be prepared and handled in accordance with the appropriate sections of the Recommended International Code of Practice - General Principles of Food Hygiene (CAC/RCP 1-1969, Rev. 3-1997) and the following relevant Codes:

(i)   the Recommended International Code of Practice for Canned Fish (CAC/RCP 10-1976);

(ii)   the Recommended International Code of Hygienic Practice for Low-Acid and Acidified Low-Acid Canned Foods (CAC/RCP 23-1979, Rev. 2-1993);

(iii)   The sections on the Products of Aquaculture in the Proposed Draft International Code of Practice for Fish and Fishery Products (under elaboration)[1]

6. LABELLING

In addition to the provisions of the Codex General Standard for Labelling of Prepackaged Foods (CODEX STAN 1-1985, Rev. 1 - 1991) the following specific provisions shall apply.

6.1 THE NAME OF THE FOOD

The name of the product shall be the designation appropriate to the species of the fish according to the law, custom or practice in the country in which the product is to be distributed.

6.2 PACKING MEDIUM

The packing medium shall form part of the name of the food.

6.3 PRESENTATION

The presentation provided for in Section 2.3.2 shall be declared in close proximity to the common name.

---

[1]   The Proposed Draft Code of Practice, when finalized, will replace all current Codes of Practice for Fish and Fishery Products

7.   SAMPLING, EXAMINATION AND ANALYSES

7.1   SAMPLING

(i)   Sampling of lots for examination of the final product as prescribed in Section 3.3 shall be in accordance with the FAO/WHO Codex Alimentarius Sampling Plans for Prepackaged Foods (AQL-6.5) (Ref. CODEX STAN 233-1969).

(ii)   Sampling of lots for examination of net weight shall be carried out in accordance with an appropriate sampling plan meeting the criteria established by the Codex Alimentarius Commission.

7.2   SENSORY EVALUATION AND PHYSICAL EXAMINATION

Samples taken for sensoric and physical examination shall be assessed by persons trained in such examination and in accordance with Section 7.3, Annex A and the *Guidelines for the Sensory Evaluation of Fish and Shellfish in Laboratories (CAC/GL 31 - 1999)*.

7.3   DETERMINATION OF NET WEIGHT

Net contents of all sample units shall be determined by the following procedure:

(i)   Weigh the unopened container.

(ii)   Open the container and remove the contents.

(iii) Weigh the empty container, (including the end) after removing excess liquid and adhering meat.

(iv)   Subtract the weight of the empty container from the weight of the unopened container. The resultant figure will be the net content.

7.4.   DETERMINATION OF DRAINED WEIGHT FOR PRODUCTS PACKED WITH EDIBLE OILS OTHER THAN SALMON OIL

The drained weight of all sample units shall be determined by the following procedure:

(i)   Maintain the container at a temperature between 20°C and 30°C for a minimum of 12 hours prior to examination.

(ii)   Open and tilt the container to distribute the contents on a pre-weighed circular sieve which consists of wire mesh with square openings of 2.8 mm x 2.8 mm.

(iii) Incline the sieve at an angle of approximately 17-20° and allow the fish to drain for two minutes, measured from the time the product is poured into the sieve.

(iv)   Weigh the sieve containing the drained fish.

(v)   The weight of drained fish is obtained by subtracting the weight of the sieve from the weight of the sieve and drained product.

8.   DEFINITION OF DEFECTIVES

A sample unit will be considered defective when it exhibits any of the properties defined below.

8.1   FOREIGN MATTER

The presence in the sample unit of any matter, which has not been derived from salmon or the packing medium does not pose a threat to human health, and is readily recognized without magnification or is present at a level determined by any method including magnification that indicates non-compliance with good manufacturing and sanitation practices.

8.2   ODOUR/FLAVOUR

A sample unit affected by persistent and distinct objectionable odours or flavours indicative of decomposition or rancidity.

8.3   TEXTURE

(i)   Excessive mushy flesh uncharacteristic of the species in the presentation; or

(ii)   Excessively tough flesh uncharacteristic of the species in the presentation; or

(iii)   Honey combed flesh in excess of 5% of the net contents.

8.4   DISCOLOURATION

A sample unit affected by distinct discolouration indicative of decomposition or rancidity or by sulphide staining of the meat exceeding 5% of the net contents.

8.5   OBJECTIONABLE MATTER

A sample unit affected by struvite crystals - any struvite crystal greater than 5 mm in length.

9.   LOT ACCEPTANCE

A lot shall be considered as meeting the requirements of this standard when:

(i)   the total number of defectives as classified according to Section 8 does not exceed the acceptance number (c) of the appropriate sampling plan in the Sampling Plans for Prepackaged Foods (AQL - 6.5) (CODEX STAN 233-1969);

(ii)   the total number of sample units not meeting the form of presentation as defined in Section 2.3 does not exceed the acceptance number (c) of the appropriate sampling plan in the Sampling Plans for Prepackaged Foods (AQL-6.5) (CODEX STAN 233-1969);

(iii)   the average net weight and the average drained weight where appropriate of all sample units examined is not less than the declared weight or drained weight as appropriate, and provided there is no unreasonable shortage in any individual container;

(iv)   the Food Additives, Hygiene and Labelling requirements of Sections 4, 5 and 6 are met.

ANNEX "A": SENSORY AND PHYSICAL EXAMINATION

1. Complete external can examination for the presence of container integrity defects or can ends which may be distorted outward.

2. Open can and complete weight determination according to defined procedures in Section 7.3 and 7.4.

3. Examine product for discolouration, foreign and objectionable matter. The presence of hard bone is an indicator of underprocessing and will require an evaluation for sterility.

4. Assess odour, flavour and texture in accordance with the *Guidelines for the Sensory Evaluation of Fish and Shellfish in Laboratories (CAC/GL 31-1999)*

CODEX STANDARD FOR CANNED TUNA AND BONITO

CODEX STAN 70 - 1981, REV.1 - 1995

1.   SCOPE

This standard applies to canned tuna and bonito. It does not apply to speciality products where the fish content constitutes less than 50% m/m of the contents.

2.   DESCRIPTION

2.1   PRODUCT DEFINITION

Canned Tuna and Bonito are the products consisting of the flesh of any of the appropriate species listed below, packed in hermetically sealed containers.

- *Thunnus alalunga*
- *Thunnus albacares*
- *Thunnus atlanticus*
- *Thunnus obesus*
- *Thunnus maccoyii*
- *Thunnus thynnus*
- *Thunnus tonggol*
- *Euthynnus affinis*
- *Euthynnus alletteratus*
- *Euthynnus lineatus*
- *Katsuwonus pelamis* (syn. *Euthynnus pelamis*)
- *Sarda chiliensis*
- *Sarda orientalis*
- *Sarda sarda*

2.2   PROCESS DEFINITION

The products shall have received a processing treatment sufficient to ensure commercial sterility.

2.3   PRESENTATION

The product shall be presented as:

2.3.1   *Solid* (skin-on or skinless) - fish cut into transverse segments which are placed in the can with the planes of their transverse cut ends parallel to the ends of the can. The proportion of free flakes or chunks shall not exceed 18% of the drained weight of the container.

2.3.2   *Chunk* - pieces of fish most of which have dimensions of not less than 1.2cm in each direction and in which the original muscle structure is retained. The proportion of pieces of flesh of which the dimensions are less than 1.2 cm shall not exceed 30% of the drained weight of the container.

2.3.3   *Flake or flakes* - a mixture of particles and pieces of fish most of which have dimensions less than 1.2 cm in each direction but in which the muscular structure of the flesh is retained. The proportion of pieces of flesh of which the dimensions are less than 1.2 cm exceed 30% of the drained weight of the container.

2.3.4   *Grated or shredded* - a mixture of particles of cooked fish that have been reduced to a uniform size, in which particles are discrete and do not comprise a paste.

2.3.5   Any other presentation shall be permitted provided that it:
- is sufficiently distinctive from other forms of presentation laid down in this standard;
- meets all other requirements of this standard;
- is adequately described on the label to avoid confusing or misleading the consumer.

3.   ESSENTIAL COMPOSITION AND QUALITY FACTORS

3.1   RAW MATERIAL

The products shall be prepared from sound fish of the species in sub-section 2.1 and of a quality fit to be sold fresh for human consumption.

3.2   OTHER INGREDIENTS

The packing medium and all other ingredients used shall be of food grade quality and conform to all applicable Codex standards.

3.3   DECOMPOSITION

The products shall not contain more than 10 mg/100 g of histamine based on the average of the sample unit tested.

3.4   FINAL PRODUCT

Products shall meet the requirements of this Standard when lots examined in accordance with Section 9 comply with the provisions set out in Section 8. Products shall be examined by the methods given in Section 7.

4.   FOOD ADDITIVES

Only the use of the following additives is permitted.

Additive	Maximum level in the Final Product

Thickening or Gelling Agents
(for use in packing media only)

400	Alginic acid	GMP
401	Sodium alginate	
402	Potassium alginate	
404	Calcium alginate	
406	Agar	
407	Carrageenan and its Na, K, and NH_4 salts (including furcelleran)	
407a	Processed *Euchema* Seaweed (PES)	
410	Carob bean gum	
412	Guar gum	
413	Tragacanth gum	
415	Xanthan gum	
440	Pectins	
466	Sodium carboxymethylcellulose	

Modified Starches

1401	Acid treated starches	GMP
1402	Alkaline treated starches	
1404	Oxidized starches	
1410	Monostarch phosphate	
1412	Distarch phosphate esterified with sodium trimetaphosphate; esterified with phosphorus oxychloride	
1414	Acetylated distarch phosphate	
1413	Phosphated distarch phosphate	
1420	Starch acetate esterified with acetic anhydride	
1421	Starch acetate esterified with vinyl acetate	
1422	Acetylated distarch adipate	
1440	Hydroxypropyl starch	
1442	Hydroxypropyl starch phosphate	

Acidity Regulators

260	Acetic acid	GMP
270	Lactic acid (L-, D-, and DL-)	
330	Citric acid	
450(i)	Disodium diphosphate	10 g/kg expressed as P_2O_5, (includes natural phosphate)

Natural Flavours

Spice oils	GMP
Spice extracts	
Smoke flavours (Natural smoke solutions and extracts)	

5.   HYGIENE AND HANDLING

5.1   The final product shall be free from any foreign material that poses a threat to human health.

5.2   When tested by appropriate methods of sampling and examination as prescribed by the Codex Alimentarius Commission, the product:

(i)   shall be free from micro-organisms capable of development under normal conditions of storage;

(ii)   no sample unit shall contain histamine that exceeds 20 mg per 100 g;

(iii)   shall not contain any other substance including substances derived from microrganisms in amounts which may represent a hazard to health in accordance with standards established by the Codex Alimentarius Commission;

(iv)   shall be free from container integrity defects which may compromise the hermetic seal.

5.3   It is recommended that the product covered by the provisions of this standard be prepared and handled in accordance with the appropriate sections of the Recommended International Code of Practice - General Principles of Food Hygiene (CAC/RCP 1-1969, Rev. 3-1997) and the following relevant Codes:

(i)   the Recommended International Code of Practice for Canned Fish (CAC/RCP 10-1976);

(ii)   the Recommended International Code of Hygienic Practice for Low-Acid and Acidified Low-Acid Canned Foods (CAC/RCP 23-1979, Rev. 2-1993);

(iii)   The sections on the Products of Aquaculture in the Proposed Draft International Code of Practice for Fish and Fishery Products (under elaboration)[1]

6.   LABELLING

In addition to the provisions of the Codex General Standard for the Labelling of Prepackaged Foods (CODEX STAN 1-1985, Rev. 1-1991) the following specific provisions apply:

6.1   THE NAME OF THE FOOD

6.1.1   The name of the product as declared on the label shall be "tuna" or "bonito", and may be preceded or followed by the common or usual name of the species, both in accordance with the law and custom of the country in which the product is sold, and in a manner not to mislead the consumer.

6.1.2   The name of the product may be qualified or accompanied by a term descriptive of the colour of the product, provided that the term "white" shall be used only for *Thunnus alalunga* and the terms "light" "dark" and "blend" shall be used only in accordance with any rules of the country in which the product is sold.

---

[1]   The Proposed Draft Code of Practice, when finalized, will replace all current Codes of Practice for Fish and Fishery Products

6.2 **FORM OF PRESENTATION**

The form of presentation provided for in Section 2.3 shall be declared in close proximity to the common name.

6.2.1 The name of the packing medium shall form part of the name of the food.

7. SAMPLING, EXAMINATION AND ANALYSES

7.1 **SAMPLING**

(i) Sampling of lots for examination of the final product as prescribed in Section 3.3 shall be in accordance with the FAO/WHO Codex Alimentarius Sampling Plans for Prepackaged Foods (AQL-6.5) (Ref. CODEX STAN 233-1969);

(ii) Sampling of lots for examination of net weight and drained weight where appropriate shall be carried out in accordance with an appropriate sampling plan established by the Commission Alimentarius Commission.

7.2 **SENSORY AND PHYSICAL EXAMINATION**

Samples taken for sensory and physical examination shall be assessed by persons trained in such examination and in accordance with the procedures set out in Sections 7.3 through 7.5, Annex A and the *Guidelines for the Sensory Evaluation of Fish and Shellfish in Laboratories (CAC/GL 31 - 1999)*.

7.3 **DETERMINATION OF NET WEIGHT**

Net contents of all sample units shall be determined by the following procedure:

(i) Weigh the unopened container.

(ii) Open the container and remove the contents.

(iii) Weigh the empty container, (including the end) after removing excess liquid and adhering meat.

(iv) Subtract the weight of the empty container from the weight of the unopened container. The resultant figure will be the net content.

7.4 **DETERMINATION OF DRAINED WEIGHT**

The drained weight of all sample units shall be determined by the following procedure:

(i) Maintain the container at a temperature between 20°C and 30°C for a minimum of 12 hours prior to examination.

(ii) Open and tilt the container to distribute the contents on a pre-weighed circular sieve which consists of wire mesh with square openings of 2.8 mm x 2.8 mm.

(iii) Incline the sieve at an angle of approximately 17-20° and allow the fish to drain for two minutes, measured from the time the product is poured into the sieve.

(iv) Weigh the sieve containing the drained fish.

(v) The weight of drained fish is obtained by subtracting the weight of the sieve from the weight of the sieve and drained product.

7.5   DETERMINATION OF WASHED DRAINED WEIGHT (FOR PACKS WITH SAUCES)

(i)   Maintain the container at a temperature between 20°C and 30°C for a minimum of 12 hours prior to examination.

(ii)   Open and tilt the container and wash the covering sauce and then the full contents with hot tap water (approx. 40°C), using a wash bottle (e.g. plastic) on the tared circular sieve.

(iii)   Wash the contents of the sieve with hot water until free of adhering sauce; where necessary separate optional ingredients (spices, vegetables, fruits) with pincers. Incline the sieve at an angle of approximately 17-20° and allow the fish to drain two minutes, measured from the time the washing procedure has finished.

(iv)   Remove adhering water from the bottom of the sieve by use of paper towel. Weigh the sieve containing the washed drained fish.

(v)   The washed drained weight is obtained by subtracting the weight of the sieve from the weight of the sieve and drained product.

7.6   DETERMINATION OF PRESENTATION

The presentation of all sample units shall be determined by the following procedure.

(i)   Open the can and drain the contents, following the procedures outlined in 7.4.

(ii)   Remove and place the contents onto a tared 1.2 cm mesh screen equipped with a collecting pan.

(iii)   Separate the fish with a spatula being careful not to break the configuration of the pieces. Ensure that the smaller pieces of fish are moved to the top of a mesh opening to allow them to fall through the screen onto the collecting pan.

(iv)   Segregate the material on the pan according to flaked, grated (shredded) or paste and weigh the individual portions to establish the weight of each component.

(v)   If declared as a "chunk" pack weigh the screen with the fish retained and record the weight. Subtract the weight of the sieve from this weight to establish the weight of solid and chunk fish.

(vi)   If declared as "solid" pack remove any small pieces (chunks) from the screen and reweigh. Subtract the weight of the sieve from this weight to establish the weight of "solid" fish.

Calculations

i)   Express the weight of flaked, grated (shredded and paste) as a percentage of the total drained weight of fish.

$$\% \text{ flakes} = \frac{\text{Weight of flakes}}{\text{Total weight of drained fish}} \times 100$$

ii)   Calculate the weight of solid and chunk fish retained on the screen by difference and express as a % of the total drained weight of fish.

$$\% \text{ solid \& chunk fish} = \frac{\text{Weight of solid \& chunk fish}}{\text{Total weight of drained fish}} \times 100$$

iii)   Calculate the weight of solid fish retained on the screen by difference and express as a % of the total drained weight of the fish.

$$\% \text{ of solid fish} = \frac{\text{Weight of solid fish}}{\text{Total weight of drained fish}} \times 100$$

7.7. DETERMINATION OF HISTAMINE

AOAC 977.13

8. DEFINITION OF DEFECTIVES

A sample unit shall be considered defective when it exhibits any of the properties defined below.

8.1 FOREIGN MATTER

The presence in the sample unit of any matter, which has not been derived from fish, does not pose a threat to human health, and is readily recognized without magnification or is present at a level determined by any method including magnification that indicates non-compliance with good manufacturing practices and sanitation practices.

8.2 ODOUR/FLAVOUR

A sample unit affected by persistent and distinct objectionable odours or flavours indicative of decomposition or rancidity.

8.3 TEXTURE

(i)   Excessively mushy flesh uncharacteristic of the species in the presentation; or

(ii)   Excessively tough flesh uncharacteristic of the species in the presentation; or

(iii)   Honey-combed flesh in excess of 5% of the drained contents.

8.4 DISCOLOURATION

A sample unit affected by distinct discolouration indicative of decomposition or rancidity or by sulphide staining of the meat exceeding 5% of the drained contents.

8.5 OBJECTIONABLE MATTER

A sample unit affected by struvite crystals greater than 5 mm in length.

9. LOT ACCEPTANCE

A lot shall be considered as meeting the requirements of this standard when:

(i)   the total number of defectives as classified according to Section 8 does not exceed the acceptance number (c) of the appropriate sampling plan in the Sampling Plans for Prepackaged Foods (AQL-6.5) (CODEX STAN 233-1969);

(ii)   the total number of sample units not meeting the presentation and colour designation as defined in Section 2.3 does not exceed the acceptance number (c) of the appropriate sampling plan in the Sampling Plans for Prepackaged Foods (AQL-6.5) (CODEX STAN 233-1969);

(iii)   the average net weight or the average weight of drained meat of all sample units examined is not less than the declared weight, and provided there is no unreasonable shortage in any individual container;

(iv)   the Food Additives, Hygiene and Labelling requirements of Sections 4, 5 and 6 are met.

ANNEX "A": SENSORY AND PHYSICAL EXAMINATION

1. Complete examination of the can exterior for the presence of container integrity defects or can ends which may be distorted outwards.
2. Open can and complete weight determination according to defined procedures in Sections 7.3 and 7.4.
3. Examine the product for discolouration.
4. Carefully remove the product and determine the presentation according to the defined procedures in Section 7.5.
5. Examine product for discolouration, foreign matter and struvite crystals. The presence of a hard bone is an indicator of under processing and will require an evaluation for sterility.
6. Assess odour, flavour and texture in accordance with the Guidelines for the Sensory Evaluation of Fish and Shellfish in Laboratories (CAC/GL 31-1999).

CODEX STANDARD FOR CANNED SARDINES AND SARDINE-TYPE PRODUCTS

CODEX STAN 94 – 1981, REV. 1 - 1995

1. SCOPE

This standard applies to canned sardines and sardine-type products packed in water or oil or other suitable packing medium. It does not apply to speciality products where fish content constitute less than 50% m/m of the net contents of the can.

2. DESCRIPTION

2.1 PRODUCT DEFINITION

2.1.1 Canned sardines or sardine type products are prepared from fresh or frozen fish of the following species:

- *Sardina pilchardus*

- *Sardinops melanostictus, S. neopilchardus, S. ocellatus, S. sagax, S. caeruleus,*

- *Sardinella aurita, S. brasiliensis, S. maderensis, S. longiceps, S. gibbosa*

- *Clupea harengus*

- *Sprattus sprattus*

- *Hyperlophus vittatus*

- *Nematalosa vlaminghi*

- *Etrumeus teres*

- *Ethmidium maculatum*

- *Engraulis anchoita, E. mordax, E. ringens*

- *Opisthonema oglinum*

2.1.2 Head and gills shall be completely removed; scales and/or tail may be removed. The fish may be eviscerated. If eviscerated, it shall be practically free from visceral parts other than roe, milt or kidney. If ungutted, it shall be practically free from undigested feed or used feed.

2.2 PROCESS DEFINITION

The products are packed in hermetically sealed containers and shall have received a processing treatment sufficient to ensure commercial sterility.

2.3 PRESENTATION

Any presentation of the product shall be permitted provided that it:

(i) contains at least two fish in each can; and

(ii) meets all requirements of this standard; and

(iii) is adequately described on the label to avoid confusing or misleading the consumer;

(iv) contain only one fish species.

3.   ESSENTIAL COMPOSITION AND QUALITY FACTORS

3.1   RAW MATERIAL

The products shall be prepared from sound fish of the species listed under sub-section 2.1 which are of a quality fit to be sold fresh for human consumption.

3.2   OTHER INGREDIENTS

The packing medium and all other ingredients used shall be of food grade quality and conform to all applicable Codex standards.

3.3.   DECOMPOSITION

The products shall not contain more than 10 mg/100 g of histamine based on the average of the sample unit tested.

3.4   FINAL PRODUCT

Products shall meet the requirements of this Standard when lots examined in accordance with Section 9 comply with provisions set out in Section 8. Product shall be examined by the methods given in Section 7.

4.   FOOD ADDITIVES

Only the use of the following additives is permitted.

Additive	Maximum Level in the Final Product
Thickening or Gelling Agents (for use in packing media only)	
400 Alginic acid	GMP
401 Sodium alginate	
402 Potassium alginate	
404 Calcium alginate	
406 Agar	
407 Carrageenan and its Na, K, and NH_4 salts (including furcelleran)	
407a Processed *Euchema* Seaweed (PES)	
410 Carob bean gum	
412 Guar gum	
413 Tragacanth gum	
415 Xanthan gum	
440 Pectins	
466 Sodium carboxymethylcellulose	
Modified Starches	
1401 Acid treated starches	GMP
1402 Alkaline treated starches	
1404 Oxidized starches	
1410 Monostarch phosphate	
1412 Distarch phosphate esterified with sodium trimetaphosphate; esterified with phosphorus oxychloride	
1414 Acetylated distarch phosphate	
1413 Phosphated distarch phosphate	
1420 Starch acetate esterified with acetic anhydride	

1421   Starch acetate esterified with vinyl acetate
1422   Acetylated distarch adipate
1440   Hydroxypropyl starch
1442   Hydroxypropyl starch phosphate

Acidity Regulators

260   Acetic acid   GMP
270   Lactic acid (L-, D-, and DL-)
330   Citric acid

Natural Flavours

Spice oils   GMP
Spice extracts
Smoke flavours (Natural smoke solutions and extracts)

5.   HYGIENE AND HANDLING

5.1   The final product shall be free from any foreign material that poses a threat to human health.

5.2   When tested by appropriate methods of sampling and examination as prescribed by the Codex Alimentarius Commission, the product:

> (i)   shall be free from microorganisms capable of development under normal conditions of storage;

> (ii)   no sample unit shall contain histamine that exceeds 20 mg per 100 g;

> (iii)   shall not contain any other substance including substances derived from microorganisms in amounts which may represent a hazard to health in accordance with standards established by the Codex Alimentarius Commission;

> (iv)   shall be free from container integrity defects which may compromise the hermetic seal.

5.3   It is recommended that the product covered by the provisions of this standard be prepared and handled in accordance with the appropriate sections of the Recommended International Code of Practice - General Principles of Food Hygiene (CAC/RCP 1-1969, Rev. 3-1997) and the following relevant Codes:

> (i)   the Recommended International Code of Practice for Canned Fish (CAC/RCP 10-1976);

> (ii)   the Recommended International Code of Hygienic Practice for Low-Acid and Acidified Low-Acid Canned Foods (CAC/RCP 23-1979, Rev. 2-1993);

6.   LABELLING

In addition to the provisions of the Codex General Standard for the Labelling of Prepackaged Foods (CODEX STAN 1-1985, Rev. 1-1991) the following specific provisions apply:

6.1   NAME OF THE FOOD

The name of the product shall be:

6.1.1   (i)   "Sardines" (to be reserved exclusively for *Sardina pilchardus* (Walbaum)); or

          (ii)   "X sardines" where "X" is the name of a country, a geographic area, the species, or the
                  common name of the species in accordance with the law and custom of the country in
                  which the product is sold, and in a manner not to mislead the consumer.

6.1.2   The name of the packing medium shall form part of the name of the food.

6.1.3   If the fish has been smoked or smoke flavoured, this information shall appear on the label in close
proximity to the name.

6.1.4   In addition, the label shall include other descriptive terms that will avoid misleading or confusing
the consumer.

7.   SAMPLING, EXAMINATION AND ANALYSES

7.1   SAMPLING

          (i)   Sampling of lots for examination of the final product as prescribed in Section 3.3 shall be in
                  accordance with the FAO/WHO Codex Alimentarius Sampling Plans for Prepackaged Foods
                  (AQL-6.5) (Ref. CODEX STAN 233-1969);

          (ii)   Sampling of lots for examination of net weight and drained weight where appropriate shall
                  be carried out in accordance with an appropriate sampling plan meeting the criteria
                  established by the Codex Alimentarius Commission.

7.2   SENSORIC AND PHYSICAL EXAMINATION

Samples taken for sensoric and physical examination shall be assessed by persons trained in such
examination and in accordance with Annex A and the *Guidelines for the Sensory Evaluation of Fish and
Shellfish in Laboratories (CAC/GL 31 - 1999).*

7.3   DETERMINATION OF NET WEIGHT

Net contents of all sample units shall be determined by the following procedure:

          (i)   Weigh the unopened container.

          (ii)   Open the container and remove the contents.

          (iii)   Weigh the empty container, (including the end) after removing excess liquid and adhering
                  meat.

          (iv)   Subtract the weight of the empty container from the weight of the unopened container.
                  The resultant figure will be the net content.

7.4   DETERMINATION OF DRAINED WEIGHT

The drained weight of all sample units shall be determined by the following procedure:

          (i)   Maintain the container at a temperature between 20°C and 30°C for a minimum of 12
                  hours prior to examination.

          (ii)   Open and tilt the container to distribute the contents on a pre-weighed circular sieve which
                  consists of wire mesh with square openings of 2.8 mm x 2.8 mm.

(iii)   Incline the sieve at an angle of approximately 17-20° and allow the fish to drain for two minutes, measured from the time the product is poured into the sieve.

(iv)   Weigh the sieve containing the drained fish.

(v)   The weight of drained fish is obtained by subtracting the weight of the sieve from the weight of the sieve and drained product.

7.5   PROCEDURE FOR PACKS IN SAUCES (WASHED DRAINED WEIGHT)

(i)   Maintain the container at a temperature between 20°C and 30°C for a minimum of 12 hours prior to examination.

(ii)   Open and tilt the container and wash the covering sauce and then the full contents with hot tap water (approx. 40°C), using a wash bottle (e.g. plastic) on the tared circular sieve.

(iii)   Wash the contents of the sieve with hot water until free of adhering sauce; where necessary separate optional ingredients (spices, vegetables, fruits) with pincers. Incline the sieve at an angle of approximately 17-20° and allow the fish to drain two minutes, measured from the time the washing procedure has finished.

(iv)   Remove adhering water from the bottom of the sieve by use of paper towel. Weigh the sieve containing the washed drained fish.

(v)   The washed drained weight is obtained by subtracting the weight of the sieve from the weight of the sieve and drained product.

7.6   DETERMINATION OF HISTAMINE

AOAC 977.13

8.   DEFINITION OF DEFECTIVES

A sample unit will be considered defective when it exhibits any of the properties defined below.

8.1   FOREIGN MATTER

The presence in the sample unit of any matter, which has not been derived from the fish or the packing media, does not pose a threat to human health, and is readily recognized without magnification or is present at a level determined by any method including magnification that indicates non-compliance with good manufacturing and sanitation practices.

8.2   ODOUR/FLAVOUR

A sample unit affected by persistent and distinct objectionable odours or flavours indicative of decomposition or rancidity.

8.3   TEXTURE

(i)   Excessively mushy flesh uncharacteristic of the species in the presentation.

(ii)   Excessively tough or fibrous flesh uncharacteristic of the species in the presentation.

8.4   DISCOLOURATION

A sample unit affected by distinct discolouration indicative of decomposition or rancidity or by sulphide staining of more than 5% of the fish by weight in the sample unit.

8.5   OBJECTIONABLE MATTER

A sample unit affected by Struvite crystals - any struvite crystal greater than 5 mm in length.

9.   LOT ACCEPTANCE

A lot will be considered as meeting the requirements of this standard when:

(i)   the total number of defectives as classified according to section 8 does not exceed the acceptance number (c) of the appropriate sampling plan in the Sampling Plans for Prepackaged Foods (AQL-6.5) (CODEX STAN 233-1969);

(ii)   the total number of sample units not meeting the presentation defined in 2.3 does not exceed the acceptance number (c) of the appropriate sampling plan in the Sampling Plans for Prepackaged Foods (AQL-6.5) (CODEX STAN 233-1969);

(iii)   the average net weight or the average drained weight where appropriate of all sample units examined is not less than the declared weight, and provided there is no unreasonable shortage in any individual container;

(iv)   the Food Additives, Hygiene and Labelling requirements of Sections 4, 5 and 6 are met.

ANNEX "A": SENSORY AND PHYSICAL EXAMINATION

1. Complete external can examination for the presence of container integrity defects or can ends which may be distorted outwards.

2. Open can and complete weight determination according to defined procedures in Sections 7.3, 7.4 and 7.5.

3. Carefully remove product and examine for discolouration, foreign matter and struvite crystals. The presence of a hard bone is an indicator of underprocessing and will require an evaluation for sterility.

4. Assess odour, flavour and texture in accordance with the *Guidelines for the Sensory Evaluation of Fish and Shellfish in Laboratories (CAC/GL 31-1999)*

CODEX STANDARD FOR CANNED SHRIMPS OR PRAWNS

CODEX STAN 37 - 1981, REV. 1 - 1995

1. SCOPE

This standard applies to canned shrimps or canned prawns.[1] It does not apply to specialty products where shrimp constitutes less than 50% m/m of the contents.

2. DESCRIPTION

2.1 PRODUCT DEFINITION

Canned shrimp is the product prepared from any combination of species of the families *Penaeidae*, *Pandalidae*, *Crangonidae* and *Palaemonidae* from which heads, shell, antennae have been removed.

2.2 PROCESS DEFINITION

Canned shrimp are packed in hermetically sealed containers and shall have received a processing treatment sufficient to ensure commercial sterility.

2.3 PRESENTATION

The product shall be presented as:

2.3.1   Peeled shrimp - shrimp which have been headed and peeled without removal of the dorsal tract;

2.3.2   Cleaned or de-veined - peeled shrimp which have had the back cut open and the dorsal tract removed at least up to the last segment next to the tail. The portion of the cleaned or de-veined shrimp shall make up 95% of the shrimp contents;

2.3.3   Broken shrimp - more than 10% of the shrimp contents consist of pieces of peeled shrimp of less than four segments with or without the vein removed;

2.3.4 Other Forms of Presentation

Any other presentation shall be permitted provided that it:

2.3.4.1 is sufficiently distinctive from other forms of presentation laid down in this standard;

2.3.4.2 meets all other requirements of this standard;

2.3.4.3 is adequately described on the label to avoid confusing or misleading the consumer.

2.3.5 Size

Canned shrimp may be designated as to size in accordance with:

     (i)   the actual count range may be declared on the label; or

     (ii)   provisions given in Annex "B".

3. ESSENTIAL COMPOSITION AND QUALITY FACTORS

3.1 SHRIMP

Shrimp shall be prepared from sound shrimp of the species in sub-section 2.1 which are of a quality fit to be sold fresh for human consumption.

---

[1]   Hereafter referred to as "shrimp".

3.2 **OTHER INGREDIENTS**

The packing medium and all other ingredients used shall be of food grade quality and conform to all applicable Codex standards.

3.3 **FINAL PRODUCT**

Products shall meet the requirements of this Standard when lots examined in accordance with Section 9 comply with the provisions set out in Section 8. Products shall be examined by the methods given in Section 7.

4. **FOOD ADDITIVES**

Only the use of the following additives is permitted.

Additive	Maximum Level in the Final Product

Colours

The following colours may be added at the level provided for in the standard for the purpose of restoring colour lost in processing:

102	Tartrazine	30 mg/kg in the
110	Sunset Yellow FCF	final product, singly
123	Amaranth	or in combination
124	Ponceau 4R	

Sequestrant

385	Calcium disodium EDTA	250 mg/kg

Acidity Regulator

330	Citric acid	GMP
338	Orthophosphoric acid	850 mg/kg

5. **HYGIENE AND HANDLING**

5.1   The final product shall be free from any foreign material, that poses a threat to human health.

5.2   When tested by appropriate methods of sampling and examination by the Codex Alimentarius Commission, the product:

(i)   shall be free from microorganisms capable of development under normal conditions of storage; and

(ii)   shall not contain any other substances including substances derived from micro organisms in amounts which may represent a hazard to health in accordance with standards established by the Codex Alimentaius Commission; and

(iii)   shall be free from container integrity defects which may compromise the hermetic seal.

5.3   It is recommended that the products covered by the provisions of this standard be prepared and handled in accordance with the appropriate sections of the Recommended International Code of Practice - General Principles of Food Hygiene (CAC/RCP 1-1969, Rev. 3-1997) and the following relevant Codes:

(i)   the Recommended International Code of Practice for Canned Fish (CAC/RCP 10-1976);

(ii)   the Recommended International Code of Hygienic Practice for Low-Acid and Acidified Low-Acid Canned Foods (CAC/RCP 23-1979, Rev. 2-1993);

(iii)   the Recommended International Code of Practice for Shrimps or Prawns (CAC/RCP 17-1978);

(iv)   The sections on the Products of Aquaculture in the Proposed Draft International Code of Practice for Fish and Fishery Products (under elaboration)[2]

6.   LABELLING

In addition to provisions of the Codex General Standard for the Labelling of Prepackaged Foods (CODEX STAN 1-1985, Rev. 1-1991) the following specific provisions apply:

6.1   THE NAME OF THE FOOD

6.1.1   The name of the product as declared on the label shall be "shrimp", or "prawns", and may be preceded or followed by the common or usual name of the species in accordance with the law and custom of the country in which the product is sold and in a manner not to mislead the consumer.

6.1.2   The name of the product shall be qualified by a term descriptive of the presentation in accordance with Sections 2.3.1 to 2.3.4.

6.1.3   If the canned shrimp are labelled as to size, the size shall comply with the provisions of Section 2.3.5 and Annex "B".

6.1.4   Broken shrimp defined in 2.3.3 shall be so labelled.

7.   SAMPLING, EXAMINATION AND ANALYSES

7.1   SAMPLING

(i)   Sampling of lots for examination of the final product as prescribed in Section 3.3 shall be in accordance with the FAO/WHO Codex Alimentarius Sampling Plans for Prepackaged Foods (AQL-6.5) (Ref. CODEX STAN 233-1969).

(ii)   Sampling of lots for examination of net weight and drained weight shall be carried out in accordance with an appropriate sampling plan meeting the criteria established by the Codex Alimentarius Commission.

7.2   SENSORIC AND PHYSICAL EXAMINATION

Samples taken for sensoric and physical examination shall be assessed by persons trained in such examination in accordance with Annex A and the *Guidelines for the Sensory Evaluation of Fish and Shellfish in Laboratories (CAC/GL 31 - 1999)*.

7.3   DETERMINATION OF NET WEIGHT

Net contents of all sample units shall be determined by the following procedure:

(i)   Weigh the unopened container;

(ii)   Open the container and remove the contents;

---

[2]   The Proposed Draft Code of Practice, when finalized, will replace all current Codes of Practice for Fish and Fishery Products

(iii)   Weigh the empty container, (including the end) after removing excess liquid and adhering meat;

(iv)   Subtract the weight of the empty container from the weight of the unopened container.

The resultant figure will be the net content.

7.4   DETERMINATION OF DRAINED WEIGHT

The drained weight of all sample units shall be determined by the following procedure:

(i)   Maintain the container at a temperature between 20°C and 30°C for a minimum of 12 hours prior to examination;

(ii)   Open and tilt the container to distribute the contents on a pre-weighed circular sieve which consists of wire mesh with square openings of 2.8 mm x 2.8 mm;

(iii)   Incline the sieve at an angle of approximately 17-20° and allow the shrimps to drain for two minutes, measured from the time the product is poured into the sieve;

(iv)   Weigh the sieve containing the drained shrimps;

(v)   The weight of drained shrimps is obtained by subtracting the weight of the sieve from the weight of the sieve and drained product.

7.5   DETERMINATION OF SIZE DESIGNATION

The size, expressed as the number of shrimp per 100g of drained product, is determined by the following equation:

$$\frac{\text{Number of whole shrimp in unit}}{\text{Actual drained weight of unit}} \times 100 = \text{Number of shrimp}/100\text{g}$$

8.   DEFINITION OF DEFECTIVES

A sample unit will be considered defective when it fails to meet any of the following final product requirements referred to in Section 3.3.

8.1   FOREIGN MATTER

The presence in the sample unit of any matter, which has not been derived from shrimp, does not pose a threat to human health, and is readily recognized without magnification or is present at a level determined by any method including magnification that indicates non-compliance with good manufacturing or sanitation practices.

8.2   ODOUR/FLAVOUR

A sample unit affected by persistent and distinct objectionable odours or flavours indicative of decomposition or rancidity.

8.3   TEXTURE

(i)   Excessive mushy flesh uncharacteristic of the species in the presentation; or

(ii)   Excessively tough flesh uncharacteristic of the species in the presentation.

8.4   DISCOLOURATION

A sample unit affected by distinct blackening of more than 10% of the surface area of individual shrimp which affects more than 15% of the number of shrimp in the sample unit.

8.5 **OBJECTIONABLE MATTER**

A sample unit affected by:

(i)   struvite crystals - any struvite crystal greater than 5 mm in length.

9. **LOT ACCEPTANCE**

A lot shall be considered as meeting the requirements of this standard when:

(i)   the total number of defectives as classified according to Section 8 does not exceed the acceptance number (c) of the appropriate sampling plan in the Sampling Plans for Prepackaged Foods (AQL-6.5) (CODEX STAN 233-1969);

(ii)   the total number of sample units not meeting presentation requirements in Section 2.3 does not exceed the acceptance number (c) of the appropriate sampling plan in the Sampling Plans for Prepackaged Foods (AQL-6.5) (CODEX STAN 233-1969);

(iii) the average net weight and the average drained weight of all sample units examined is not less than the declared weight and provided there is no unreasonable shortage in any individual container;

(iv)   the Food Additives, Hygiene and Labelling requirements of Sections 4, 5 and 6 are met.

ANNEX "A": SENSORY AND PHYSICAL EXAMINATION

1. Complete external can examination for the presence of container integrity defects or can ends which may be distorted outwards.
2. Open can and complete weight determination according to defined procedures in Sections 7.3 and 7.4.
3. Carefully remove the product and examine for size designation in accordance with the procedure in Section 7.5.
4. Examine product for discolouration, foreign and objectionable matter.
5. Assess odour, flavour and texture in accordance with the Guidelines for the Sensory Evaluation of Fish and Shellfish in Laboratories (CAC/GL 31-1999)

ANNEX "B": SIZE DESIGNATION OF CANNED SHRIMPS

The terms "extra large", "jumbo", "large", "medium", "small", "tiny" may be used provided that the range is in accordance with the following table:

Number of whole shrimp (including pieces greater than 4 segments) per 100g of drained product

SIZE DESIGNATION	RANGE
Extra Large or Jumbo	13 or less
Large	14-19
Medium	20-34
Small	35-65
Tiny	more than 65

CODEX STANDARD FOR CANNED CRAB MEAT

CODEX STAN 90 - 1981, REV.1 - 1995

1.   SCOPE

This standard applies to canned crab meat. It does not apply to specialty products where crab meat constitutes less than 50% m/m of the contents.

2.   DESCRIPTION

2.1   PRODUCT DEFINITION

Canned crab meat is prepared singly or in combination from the leg, claw, body and shoulder meat from which the shell has been removed, of any of the edible species of the sub-order *Brachyura* of the order Decapoda and all species of the family *Lithodidae*.

2.2   PROCESS DEFINITION

Canned crab meat is packed in hermetically sealed containers and shall have received a processing treatment sufficient to ensure commercial sterility.

2.3   PRESENTATION

Any presentation of the product shall be permitted provided that it:

(i)   meets all requirements of this standard; and

(ii)   is adequately described on the label to avoid confusing or misleading the consumer.

3.   ESSENTIAL COMPOSITION AND QUALITY FACTORS

3.1   CRAB MEAT

Canned crab meat shall be prepared from sound crab of the species designated in 2.1 which are alive immediately prior to the commencement of processing and of a quality suitable for human consumption.

3.2   OTHER INGREDIENTS

The packing medium and all other ingredients used shall be of food grade quality and conform to all applicable Codex standards.

3.3   FINAL PRODUCT

Products shall meet the requirements of this Standard when lots examined in accordance with Section 9 comply with provisions set out in Section 8. Products shall be examined by the methods given in Section 7.

4.   FOOD ADDITIVES

Only the use of the following additives is permitted.

Additive	Maximum Level in the final product
Acidity Regulators	
330 Citric acid	GMP

338	Orthophosphoric acid	10 g/kg expressed as
450(i)	Disodium diphosphate	P_2O_5, singly or in combination (includes natural phosphate)

Sequestrant

385	Calcium disodium EDTA	250 mg/kg

Flavour Enhancer

621	Monosodium glutamate	GMP

5.   HYGIENE AND HANDLING

5.1   The final product shall be free from any foreign material that poses a threat to human health.

5.2   When tested by appropriate methods of sampling and examination prescribed by the Codex Alimentarius Commission, the product:

(i)   shall be free from microorganisms capable of development under normal conditions of storage; and

(ii)   shall not contain any other substance including substances derived from microorganisms in amounts which may represent a hazard to health in accordance with standards established by the Codex Alimentarius Commission; and

(iii)   shall be free from container integrity defects which may compromise the hermetic seal.

5.3   It is recommended that the product covered by the provisions of this standard be prepared and handled in accordance with the appropriate sections of the Recommended International Code of Practice - General Principles of Food Hygiene (CAC/RCP 1-1969, Rev. 3-1997) and the following relevant Codes:

(i)   the Recommended International Code of Practice for Canned Fish (CAC/RCP 10-1976);

(ii)   the Recommended International Code of Hygienic Practice for Low-Acid and Acidified Low-Acid Canned Foods (CAC/RCP 23-1979, Rev. 2-1993);

(iii)   the Recommended International Code of Practice for Crabs (CAC/RCP 28-1983);

(iv)   The sections on the Products of Aquaculture in the Proposed Draft International Code of Practice for Fish and Fishery Products (under elaboration)[1]

6.   LABELLING

In addition to provisions of the Codex General Standard for the Labelling of Prepackaged Foods (CODEX STAN 1-1985, Rev. 1-1991) the following specific provisions apply:

6.1   NAME OF THE FOOD

6.1.1   The name of the product shall be "crab" or "crabmeat".

6.1.2   In addition, the label shall include other descriptive terms that will avoid misleading or confusing the consumer.

---

[1]   The Proposed Draft Code of Practice, when finalized, will replace all current Codes of Practice for Fish and Fishery Products

7.   SAMPLING, EXAMINATION AND ANALYSES

7.1   SAMPLING

(i)   Sampling of lots for examination of the final product as prescribed in Section 3.3 shall be in accordance with the FAO/WHO Codex Alimentarius Sampling Plans for Prepackaged Foods (AQL-6.5) (Ref. CODEX STAN 233-1969).

(ii)   Sampling of lots for examination of net weight and drained weight shall be carried out in accordance with an appropriate sampling plan meeting the criteria established by the Codex Alimentarius Commission.

7.2   SENSORIC AND PHYSICAL EXAMINATION

Samples taken for sensoric and physical examination shall be assessed by persons trained in such examination and in accordance with Annex A and the *Guidelines for the Sensory Evaluation of Fish and Shellfish in Laboratories (CAC/GL 31 - 1999)*.

7.3   DETERMINATION OF NET WEIGHT

Net weight of all sample units shall be determined by the following procedures:

(i)   Weigh the unopened container.

(ii)   Open the container and remove the contents.

(iii)   Weigh the empty container, including the end and any wrapping material, after removing excess liquid and adhering meat.

(iv)   Subtract the weight of the empty container and any wrapping material from the weight of the unopened container. The resultant figure is the net content.

7.4   DETERMINATION OF DRAINED WEIGHT

The drained weight of all sample units shall be determined by the following procedures:

(i)   Maintain the container at a temperature of between 20°C and 30°C for a minimum of 12 hours prior to examination.

(ii)   Open the container and distribute the contents on a pre-weighed circular sieve having a wire mesh with square openings of 2.8 mm x 2.8 mm.

(iii)   Remove all wrapping material and incline the sieve at an angle of approximately 17-20° and allow the meat to drain two minutes, measured from the time the product is poured onto the sieve.

(iv)   Weigh the sieve containing the drained crab meat.

(v)   Determine the weight of drained crab meat by subtracting the mass of the sieve from the mass of the sieve with drained product.

8.   DEFINITION OF DEFECTIVES

A sample unit will be considered defective when it exhibits any of the properties defined below.

8.1   FOREIGN MATTER

The presence in the sample unit of any matter, which has not been derived from crab meat, does not pose a threat to human health, and is readily recognized without magnification or is present at a level determined by any method including magnification that indicates non-compliance with good manufacturing and sanitation practices.

8.2   ODOUR/FLAVOUR

A sample unit affected by persistent and distinct objectionable odours or flavours indicative of decomposition or rancidity.

8.3   TEXTURE

(i)   Excessively mushy flesh uncharacteristic of the species in the presentation; or

(ii)   Excessively tough flesh uncharacteristic of the species in the presentation.

8.4   DISCOLOURATION

A sample unit affected by distinct discolourations indicative of decomposition or rancidity or by blue, brown, black discolourations exceeding 5% by weight of the drained contents, or black sulphide staining of the meat exceeding 5% by weight of the drained contents.

8.5   OBJECTIONABLE MATTER

A sample unit affected by struvite crystals - any struvite crystal greater than 5 mm in length.

9.   LOT ACCEPTANCE

A lot shall be considered as meeting the requirements of this standard when:

(i)   the total number of defectives as classified according to Section 8 does not exceed the acceptance number (c) of the appropriate sampling plan in the Sampling Plans for Prepackaged Foods (AQL-6.5) (CODEX STAN 233-1969).

(ii)   the total number of sample units not meeting the form of presentation defined in Section 2.3 does not exceed the acceptance number (c) of the appropriate sampling plan in the Sampling Plans for Prepackaged Foods (AQL - 6.5) (CODEX STAN 233-1969);

(iii)   the average net weight and the average drained weight where appropriate of all sample units examined is not less than the declared weight, and provided there is no unreasonable shortage in any individual container.

(iv)   the Food Additives, Hygiene and Labelling requirements of Sections 4, 5 and 6 are met.

ANNEX "A": SENSORY AND PHYSICAL EXAMINATION

1. Complete external can examination for the presence of container integrity defects or can ends which may be distorted outwards.

2. Open can and complete weight determination according to defined procedures in Sections 7.3 and 7.4.

3. Examine product for discolouration, foreign and objectionable matter.

4. Assess odour, flavour and texture in accordance with the *Guidelines for the Sensory Evaluation of Fish and Shellfish in Laboratories (CAC/GL 31-1999)*

SECTION 3

OTHER FISH AND FISHERY PRODUCTS

CODEX STANDARD FOR SALTED FISH AND DRIED SALTED FISH
OF THE GADIDAE FAMILY OF FISHES

CODEX STAN 167 - 1989, Rev.1-1995

1.   SCOPE

This standard applies to salted fish and dried salted fish of the *Gadidae* family which has been fully saturated with salt (heavy salted) or to salted fish which has been preserved by partial saturation to a salt content not less than 12% by weight of the salted fish which may be offered for consumption without further industrial processing.

2.   DESCRIPTION

2.1   PRODUCT DEFINITION

Salted fish is the product obtained from fish:

(a)   of the species belonging to the family *Gadidae*; and

(b)   which has been bled, gutted, beheaded, split or filleted, washed, salted.

(c)   dried salted fish is salted fish which have been dried.

2.2   PROCESS DEFINITION

The product shall be prepared by one of the salting processes defined in 2.2.1 and one or both of the drying processes defined in 2.2.2 and according to the different types of presentation as defined in 2.3.

2.2.1   Salting

(a)   Dry Salting (kench curing) is the process of mixing fish with suitable food grade salt and stacking the fish in such a manner that the excess of the resulting brine drains away.

(b)   Wet Salting (pickling) is the process whereby fish is mixed with suitable food grade salt and stored in watertight containers under the resultant brine (pickle) which forms by solution of salt in the water extracted from the fish tissue. Brine may be added to the container. The fish is subsequently removed from the container and stacked so that the brine drains away.

(c)   Brine Injection is the process for directly injecting brine into the fish flesh and is permitted as a part of the heavy salting process.

2.2.2   Drying

(a)   Natural Drying - the fish is dried by exposure to the open air; and

(b)   Artificial Drying - the fish is dried in mechanically circulated air, the temperature and humidity of which may be controlled.

2.3   PRESENTATION

2.3.1   **Split fish** - split and with the major length of the anterior of the backbone removed (about two-thirds).

2.3.2   **Split fish with entire backbone** - split with the whole of the backbone not removed.

2.3.3   **Fillet** - is cut from the fresh fish, strips of flesh is cut parallel to the central bone of the fish and from which fins, main bones and sometimes belly flap is removed.

2.3.4   Other presentation: any other presentation of the product shall be permitted provided that it

     (i)   is sufficiently distinctive from the other forms of presentation laid down in this Standard;

     (ii)   meets all other requirements of this Standard; and

     (iii)   is adequately described on the label to avoid confusing or misleading the consumer.

2.3.5   Individual containers shall contain only one form of presentation from only one species of fish.

3.   ESSENTIAL COMPOSITION AND QUALITY FACTORS

3.1   FISH

Salted fish shall be prepared from sound and wholesome fish, fit for human consumption.

3.2   SALT

Salt used to produce salted fish shall be clean, free from foreign matter and foreign crystals, show no visible signs of contamination with dirt, oil, bilge or other extraneous materials and comply with the requirements laid down in supplement 1 to the Code of Practice for Salted Fish (CAC/RCP 26-1979).

3.3   FINAL PRODUCT

Products shall meet the requirements of this standard when lots examined in accordance with Section 9. comply with the provisions set out in Section 8. Products shall be examined by the methods given in Section 7.

4.   FOOD ADDITIVES

Only the use of following additives is permitted.

Additives	Maximum level in the Final Product
Preservatives	
200 Sorbic acid	200 mg/kg, singly
201 Sodium sorbate	or in combination
202 Potassium sorbate	expressed as sorbic acid

5.   HYGIENE AND HANDLING

5.1   The final product shall be free from any foreign material that poses a threat to human health.

5.2   When tested by appropriate methods of sampling and examination prescribed by the Codex Alimentarius Commission , the product:

     (i)   shall be free from microorganisms or substances originating from microorganisms in amounts which may present a hazard to health in accordance with standards established by the Codex Alimentarius Commission;

     (ii)   shall not contain any other substance in amounts which may present a hazard to health in accordance with standards established by the Codex Alimentarius Commission.

5.3   It is recommended that the products covered by the provisions of this standard be prepared and handled in accordance with the appropriate sections of the Recommended International Code of Practice - General Principles of Food Hygiene (CAC/RCP 1-1969, Rev. 3-1997) and the following relevant Codes:

     (i)   the Recommended International Code of Practice for Fresh Fish (CAC/RCP 9-1976);

     (ii)   the Recommended International Code of Practice for Frozen Fish (CAC/RCP 16-1978);

(iii)   the Recommended International Code of Practice for Salted Fish (CAC/RCP 26-1979).

(iv)   The sections on the Products of Aquaculture in the Proposed Draft International Code of Practice for Fish and Fishery Products (under elaboration)[1]

6.   LABELLING

In addition to the provisions of the Codex General Standard for the Labelling of Prepackaged Foods (CODEX STAN 1-1985, Rev. 1-1991), the following specific provisions apply:

6.1   THE NAME OF THE FOOD

6.1.1   The name of the food to be declared on the label shall be   "salted fish", "wet salted fish" or "salted fillet"   "dried salted fish" or "klippfish" or other designations according to the law, custom or practice in the country in which the product is to be distributed. In addition, there shall appear on the label in conjunction with the name of the product, the name of the species of fish from which the product is derived.

6.1.2   For forms of presentation other than those described in 2.3.1 "split fish", the form of presentation shall be declared in conjunction with the name of the product in accordance with sub-section 2.3.2 as appropriate. If the product is produced in accordance with sub-section 2.3.3, the label shall contain in close proximity to the name of the food, such additional words or phrases that will avoid misleading or confusing the consumer.

6.1.3   The term "klippfish" can only be used for dried salted fish which has been prepared from fish which has reached 95% salt saturation prior to drying.

6.1.4   The term "wet salted fish" can only be used for fish fully saturated with salt.

6.2   LABELLING OF NON-RETAIL CONTAINERS

Information specified above shall be given either on the container or in accompanying documents, except that the name of the food, lot identification, and the name and address of the manufacturer or packer shall always appear on the container.

However, lot identification, and the name and address may be replaced by an identification mark, provided that such a mark is clearly identifiable with the accompanying documents.

7.   SAMPLING, EXAMINATION AND ANALYSES

7.1   SAMPLING

(i)   Sampling of lots for examination of the product shall be in accordance with the FAO/WHO Codex Alimentarius Sampling Plans for Prepackaged Foods (AQL - 6.5) (CODEX STAN 233-1969). A sample unit shall be the primary container or where the product is in bulk, the individual fish is the sample unit.

(ii)   Sampling for net weight shall be carried out in accordance with the FAO/WHO Sampling Plans for the Determination of Net Weight (under elaboration).

7.2   SENSORY AND PHYSICAL EXAMINATION

Samples taken for sensory and physical examination shall be assessed by persons trained in such examination and in accordance with procedures elaborated in Annex A and in accordance with *Guidelines for the Sensory Evaluation of Fish and Shellfish in Laboratories (CAC/GL 31 - 1999).*

---

[1]   The Proposed Draft Code of Practice, when finalized, will replace all current Codes of Practice for Fish and Fishery Products

7.3   DETERMINATION OF NET WEIGHT

The net weight (excluding packaging material and excess salt) of each sample unit in the sample lot shall be determined.

7.4   DETERMINATION OF SALT CONTENT

1.   Principle

The salt is extracted by water from the preweighed sample. After the precipitation of the proteins, the chloride concentration is determined by titration of an aliquot of the solution with a standardized silver nitrate solution (Mohr method) and calculated as sodium chloride.

2.   Equipment and chemicals
- Brush
- Sharp knife or saw
- Balance, accurate to ±0.01 g
- Calibrated volumetric flasks, 250 ml
- Erlenmeyer flasks
- Electric homoginizer
- Magnetic stirrer
- Folded paper filter, quick running
- Pipettes
- Funnel
- Burette
- Potassium hexacyano ferrate (II), $K_4Fe(CN)_6 \cdot 3H_2O$, 15% w/v (aq)
- Zinc sulphate, $ZnSO_4 \cdot 6H_2O$, 30% w/v (aq)
- Sodium hydroxide, NaOH, 0.1 N, 0.41% w/v (aq)
- Silver nitrate, $AgNO_3$, 0.1 N, 1.6987% w/v (aq), standardized
- Potassium chromate, K_2CrO_4 5% w/v (aq)
- Phenolphthalein, 1% in ethanol
- distilled or deionized water

3.   Preparation of sample

Before preparing a subsample adhering salt crystals should be removed by brushing from the surface of the sample without using water.

The entire sample should be subjected to a systematic cutting and randomization process to assure a subsample representative of the composition of the whole fish or fishery product.

At least 100 g of subsample should be thoroughly homogenized by using an electric homoginizer.

Determination should be performed at least in duplicate.

4. Procedure

(i) Five gram of homoginized subsample is weighted into a 250 ml volumetric flask and vigorously shaken with approximately 100 ml water.

(ii) Five millilitre of potassium hexacyano ferrate solution and 5 ml of zinc sulphate solution are added, the flask is shaken.

(iii) Water is added to the graduation mark.

(iv) After shaking again and allowing to stand for precipitation, the flask content is filtered through a folded paper filter.

(v) An aliquot of the clear filtrate is transferred into an Erlenmeyer flask and two drops of phenolphthalein are added. Sodium hydroxide is added dropwise until the aliquot takes on a faint red colour. The aliquot then diluted with water to approximately 100 ml.

(vi) After addition of approximately 1 ml potassium chromate solution, the diluted aliquot is titrated under constant stirring, with silver nitrate solution. Endpoint is indicated by a faint, but distinct, change in colour. This faint reddish-brown colour should persist after brisk shaking.

To recognize the colour change, it is advisable to carry out the titration against a white background.

(vii) Blank titration of reagents used should be done.

(viii) Endpoint determination can also be made by using instruments like potentiometer or coulorimeter.

5. Calculation of results

In the equation of the calculation of results the following symbols are used:

A = volume of aliquot (ml)

C = concentration of silver nitrate solution in N

V = volume of silver nitrate solution in ml used to reach endpoint and corrected for blank value

W = sample weight (g)

The salt content in the sample is calculated by using the equation:

Salt concentration (%) = (V x C x 58.45 x 250 x 100) / (A x W x 1000)

Results should be reported with one figure after the decimal point.

6. Reference method

As reference method a method should be used which includes the complete ashing of the sample in a muffle furnace at 550°C before chloride determination according to the method described above (leaving out steps (ii) and (iv)).

7. Comments

By using the given equation all chloride determined is calculated as sodium chloride. However it is impossible to estimate sodium by this methodology, because other chlorides of the alkali and earth alkali elements are present which form the counterparts of chlorides.

The presence of natural halogens other than chloride in fish and salt is negligible.

A step, in which proteins are precipitated (ii), is essential to avoid misleading results.

8.   DEFINITION OF DEFECTIVES

8.1   The sample unit shall be considered defective when it exhibits any of the properties defined below.

8.1.1   Foreign Matter

The presence in the sample unit of any matter which has not been derived from Gadidae fish, does not pose a threat to human health, and is readily recognized without magnification or is present at a level determined by any method including magnification that indicates non-compliance with good manufacturing and sanitation practices.

8.1.2   Odour

A fish affected by persistent and distinct objectionable odours indicative of decomposition (such as sour, putrid, etc.) or contamination by foreign substances (such as fuel oil, cleaning compounds, etc.).

8.1.3   Pink

Any visible evidence of red halophilic bacteria.

8.1.4   Appearance

Textural breakdown of the flesh which is characterized by extensive cracks on more than 2/3 of the surface area or which has been mutilated, torn or broken through to the extent that the split fish is divided into two or more pieces but still held together by skin.

8.2   The sample unit shall be considered defective when 30% or more of the fish in the sample unit are affected by any of the following defects.

8.2.1   Halophilic Mould (dun)

A fish showing an aggregate area of pronounced halophilic mould clusters on more than 1/3 of the total surface area of the face side.

8.2.2   Liver Stains

A pronounced yellow or yellowish orange discoloration caused by the presence of liver and affecting more than 1/4 of the total surface area of the face of the fish.

8.2.3   Intense Bruising

Any fish showing more than 1/2 of the face of the fish with intense bruising.

8.2.4   Severe Burning

A fish with more than 1/2 of the back (skin side) tacky or sticky due to overheating during drying.

9.   **LOT ACCEPTANCE**

A lot shall be considered as meeting the requirements of this standard when:

(i)   the total number of defectives as classified according to section 8 does not exceed the acceptance number (c) of the appropriate sampling plan in the Sampling Plans for Prepackaged Foods (AQL-6.5) (CODEX STAN 233-1969);

(ii)   the average net weight of all sample units is not less than the declared weight, provided no individual container is less than 95% of the declared weight; and

(iii)   the total number of sample units not meeting the form of presentation as defined in section 2.3 does not exceed the acceptance number (c) of the appropriate sampling plan in the Sampling Plans for prepackaged Foods (AQL - 6.5) (CODEX STAN 233-1969);

(iv)   the Food Additives, Hygiene and Handling and Labelling requirements of Sections 4, 5 and 6 are met.

"ANNEX A": SENSORY AND PHYSICAL EXAMINATION

1. Examine every fish in the sample in its entirety.

2. Examine the product for the form of presentation.

3. Examine the fish for foreign matter, pink conditions, halophilic mould, liver stains, intense bruising, severe burning and texture.

4. Assess odour in accordance with the *Guidelines for the Sensory Evaluation of Fish and Shellfish in Laboratories (CAC/GL 31 - 1999)*.

CODEX STANDARD FOR DRIED SHARK FINS

CODEX STAN 189 - 1993

1. SCOPE

This Standard applies to dried shark fins intended for further processing.

2. DESCRIPTION

2.1 PRODUCT DEFINITION

Dried shark fins are the dorsal and pectoral fins cut in the form of an arc and the lower lobe of the caudal fin cut straight, from which all flesh has been removed, and are cut from species of sharks which are safe for human consumption.

2.2 PROCESS DEFINITION

The fins shall be subjected to a drying process so as to meet the requirements of Section 3.2.4 and shall comply with the conditions laid down hereafter.

2.3 PRESENTATION

2.3.1 Dried shark fins may be presented with the skin on or as skinless.

2.3.2 Other Forms of Presentation

Any other presentation shall be permitted provided that it:

(i)   meets all other requirements of this standard; and

(ii)   is adequately described on the label to avoid confusing or misleading the consumer.

3. ESSENTIAL COMPOSITION AND QUALITY FACTORS

3.1 SHARK

Dried shark fins shall be prepared from sound sharks which are of a quality fit to be sold fresh for human consumption.

3.2 OTHER INGREDIENTS

None.

3.3 FINAL PRODUCT

3.3.1 Appearance

The final product shall be free from foreign material.

3.3.2 Odour

The product shall be free from objectionable odours.

3.3.3 Texture

The dried shark fins shall be free from objectionable textural characteristics.

3.3.4 Percentage of Moisture

The final product shall have a moisture content not exceeding 18%.

4. FOOD ADDITIVES

No additives are permitted.

5. HYGIENE AND HANDLING

5.1   The final product shall be free from any foreign material that poses a threat to human health.

5.2   When tested by appropriate methods of sampling and examination prescribed by the Codex Alimentarius Commission, the product:

(i)   shall be free from microorganisms or substances originating from microorganisms in amounts which may present a hazard to health in accordance with standards established by the Codex Alimentarius Commission;

(ii)   shall not contain any other substance in amounts which may present a hazard to health in accordance with standards established by the Codex Alimentarius Commission.

5.3   It is recommended that the product covered by the provisions of this standard be prepared and handled in accordance with the appropriate sections of the Recommended International Code of Practice - General Principles of Food Hygiene (CAC/RCP 1-1969, Rev. 3-1997) and the following relevant Code: Recommended International Code of Practice for Fresh Fish (CAC/RCP 9-1976).

6. LABELLING

In addition to the General Standard for the Labelling of Prepackaged Foods (CODEX STAN 1-1985, Rev. 1-1991), the following specific provisions shall apply:

6.1 NAME OF THE FOOD

The name of the product shall be "dried shark fins" or any other appropriate name in accordance with the law and custom of the country in which the product is to be distributed.

6.1.1   There shall appear on the label reference to the form of presentation in close proximity to the name of the product in such descriptive terms that will adequately and fully describe the nature of the presentation of the product to avoid misleading or confusing the consumer.

6.1.2   In addition to the specified labelling designations above, the name of the species, the type of fin, and its size shall also appear on the label.

6.2 LABELLING OF NON-RETAIL CONTAINERS

Information on the above provisions shall be given either on the container or in accompanying documents, except that the name of the product, lot identification, and the name and address of the manufacturer or packer, shall appear on the container.

However, lot identification, and the name and address of the manufacturer or packer may be replaced by an identification mark provided that such a mark is clearly identifiable with the accompanying documents.

7. SAMPLING, EXAMINATION AND ANALYSIS

7.1 SAMPLING

(i)   Sampling of lots for examination of the product shall be in accordance with the Codex Sampling Plans of Prepackaged Foods (AQL-6.5) (CODEX STAN 233-1969);

(ii)   The sampling of lots for examination of net weight shall be carried out according to the Codex Sampling Plans for the Determination of Net Weight (under elaboration).

7.2   SENSORY AND PHYSICAL EXAMINATION

Samples taken for sensory and physical examination shall be assessed by persons trained in such examination and in accordance with the procedures set out in Section 7.3, Annex B "Sensoric and Physical Examination"[1] and the *Guidelines for the Sensory Evaluation of Fish and Shellfish in Laboratories (CAC/GL 31 - 1999)*.

7.3   DETERMINATION OF NET WEIGHT

The net weight (exclusive of packaging material) of each sample unit in the sample lot shall be determined.

7.4   DETERMINATION OF MOISTURE

[Method to be developed.]

8.   CLASSIFICATION OF DEFECTIVES

A sample unit shall be considered defective when it fails to meet any of the following final product requirements referred to in Section 3.3.

8.1   FOREIGN MATTER

The presence in the sample unit of any matter which has not been derived from fish, does not pose a threat to human health, and is readily recognized without magnification or is present at a level determined by any method including magnification that indicates non-compliance with good manufacturing and sanitation practices.

8.2   ODOUR

A sample unit affected by persistent and distinct objectionable odours indicative of decomposition.

8.3   TEXTURE

Textural breakdown of the fin, indicative of decomposition, characterized by softness.

8.4   MOISTURE

The sample unit exceeds 18% moisture.

9.   LOT ACCEPTANCE

A lot shall be considered as meeting the requirements of this Standard when:

(i)   the total number of defectives as classified according to Section 8 does not exceed the acceptance number (c) of the appropriate sampling plan in the Sampling Plans for Prepackaged Foods (AQL-6.5) (CODEX STAN 233-1969);

(ii)   the average net weight of all sample units is not less than the declared weight, provided there is no unreasonable shortage in any container; and

(iii)   the total number of sample units not meeting the form of presentation as defined in Section 2.3 does not exceed the acceptance number (c) of the appropriate sampling plan in the Sampling Plans for prepackaged Foods (AQL - 6.5) (CODEX STAN 233-1969);

(iv)   the Food Additive, Hygiene and Handling and Labelling requirements of Sections 4, 5 and 6 are met.

---

[1]   To be developed.

ANNEX A

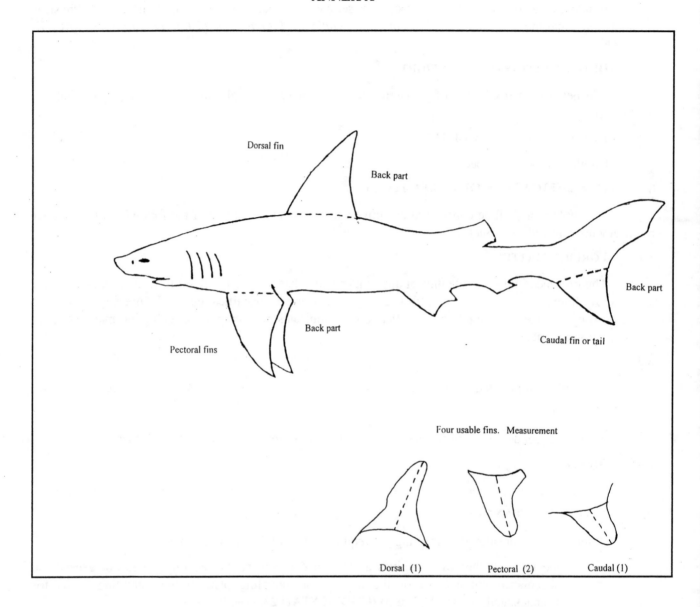

STANDARD FOR CRACKERS FROM MARINE AND FRESHWATER FISH, CRUSTACEAN AND MOLLUSCAN SHELLFISH

CODEX STAN 222-2001

1. SCOPE

This standard shall apply to crackers prepared from marine and freshwater fish, crustacean and molluscan shellfish. It does not include ready-to-eat fried as well as artificially flavoured fish, crustacean and molluscan shellfish crackers.

2. DESCRIPTION

2.1 PRODUCT DEFINITION

The product is a traditional food made from fresh fish or frozen minced flesh of either marine (including both the red meat and white meat species) or freshwater fish, crustacean (including prawns and shrimps) and molluscan shellfish (including squids, cuttlefish, oysters, clams, mussels and cockles) as described in section 3.1 and other ingredients as described in section 3.2.

2.2 PROCESS DEFINITION

2.2.1 The product shall be prepared by mixing all the ingredients, forming, cooking, cooling, slicing and drying.

2.2.2 The product shall be packed in a suitable packaging material which is moisture proof and gas impermeable. It shall be processed and packaged so as to minimize oxidation.

2.3 HANDLING PRACTICE

Fresh marine and freshwater fish, crustacean and molluscan shellfish shall be preserved immediately after harvesting by chilling or icing to bring its temperature down to 0°C (32°F) as quickly as possible as specified in the Recommended International Code of Practice for Fresh Fish (CAC/RCP 9-1976) and kept at an adequate temperature to prevent spoilage and bacterial growth prior to processing.

3. ESSENTIAL COMPOSITION AND QUALITY FACTORS

3.1 RAW MATERIAL

Fresh marine and freshwater fish, crustacean and molluscan shellfish shall mean freshly caught, chilled or frozen marine and freshwater fish, crustacean and molluscan shellfish. Frozen minced flesh shall mean freshly caught, chilled or frozen marine and freshwater fish, crustacean and molluscan shellfish which has been appropriately processed. The marine and freshwater fish, crustacean and molluscan shellfish shall have a characteristic fresh appearance, colour and odour.

3.2 OTHER INGREDIENTS

Other ingredients shall be of food grade quality and conform to all applicable Codex Standards.

3.3 OPTIONAL INGREDIENTS

The product may contain sugar as well as suitable spices.

3.4 FINAL PRODUCT

3.4.1 The product shall display a uniform size, shape, colour, thickness and texture.

3.4.2 The product shall comply with the requirements prescribed in Table 1.

TABLE 1 :   REQUIREMENTS FOR CRACKERS FROM MARINE AND FRESHWATER FISH, CRUSTACEAN AND MOLLUSCAN SHELLFISH

Characteristics	Grade	Fish	Crustacean and Molluscan Shellfish
Crude protein (N x 6.25), percent w/w min.	I	12	8
	II	8	5
	III	5	2
Moisture content, percent w/w	I	)	)
	II	) 8 to 14	) 8 to 14
	III	)	)

4.   FOOD ADDITIVES

Additives
Sequestrants

Maximum Level in the Final Product

452   Polyphosphates   5g/kg expressed as P_2O_5, single or in combination

Flavour enhancers

621   Monosodium glutamate   Limited by GMP

5.   HYGIENE

5.1   It is recommended that the product covered by the provisions of this standard be prepared and handled in accordance with the appropriate sections of the Recommended International Code of Practice - General Principles of Food Hygiene (CAC/RCP 1-1985, Rev 2-1997), and the Recommended International Code of Practice for Fresh Fish (CAC/RCP 9 - 1976).

5.2   The products should comply with any microbiological criteria established in accordance with the Principles for the Establishment and Application of Microbiological Criteria for Foods (CAC/GL 21-1997).

6.   LABELLING

In addition to the provisions of the Codex General Standard for the Labelling of Prepackaged Foods (CODEX STAN 1-1985, Rev.1 -1991), the following specific provisions apply:

6.1   THE NAME OF THE FOOD

The name of the product from marine and freshwater fish shall be "Fish Crackers" and those from crustacean and molluscan shellfish shall depict the common name of the species, like "Prawn Crackers" or "Squid Crackers".

6.2   GRADES

When declared by grade, the package shall declare the grade as prescribed in Table 1.

6.3   ADDITIONAL REQUIREMENTS

The package shall bear clear directions for keeping the product from the time it is purchased from the retailer to the time of its use and directions for cooking.

7. SAMPLING, EXAMINATION AND ANALYSIS

7.1 SAMPLING

Sampling of lots for examination of the products shall be in accordance with the FAO/WHO Codex Alimentarius Sampling Plans for Prepackaged Foods (1969) (AQL-6.5) (CODEX STAN 233-1969).

7.2 DETERMINATION OF CRUDE PROTEIN

According to AOAC 920.87 or 960.52.

7.3 DETERMINATION OF MOISTURE

According to AOAC 950.46B (air drying).

7.4 SENSORY AND PHYSICAL EXAMINATION

Samples taken for sensory and physical examination shall be assessed by persons trained in such examination and in accordance with Annex A.

8. DEFINITION OF DEFECTIVES

The sample unit shall be considered defective when it exhibits any of the properties defined below:

8.1 FOREIGN MATTER

The presence in the sample unit of any matter which has not been derived from materials specified in section 3.1, 3.2, 3.3, does not pose a threat to human health and is readily recognized without magnification that indicates non-compliance with good manufacturing and sanitation practices.

8.2 ODOUR AND FLAVOUR

Unfried crackers affected by persistent and distinct objectionable odours and fried crackers affected by persistent and distinct objectionable flavours indicative of decomposition (such as putrid), or contamination by foreign substances (such as fuel oil and cleaning compound).

8.3 BONES

Crackers with more than one bone greater than 3 mm in diameter and 5mm in length that affects more than 25% of the sample unit.

8.4 DISCOLOURATION

Pronounced black, whitish or yellowish discolouration indicative of mould or fungal growth on the surface of crackers that affects more than 10% of the sample unit.

9. LOT ACCEPTANCE

A lot shall be considered as meeting the requirements of this standard when:

1. the total number of defectives as classified according to Section 8 does not exceed the acceptable number of the appropriate sampling plan in the Sampling Plans for Prepackaged Foods (1969) (AQL-6.5) (CODEX STAN 233-1969).

2. the average net weight of all sample units is not less than the declared weight, provided no individual container is less than 95% of the declared weight; and

3. the Food Additives, Hygiene, Packing and Labelling requirements of Section 4, 5, 2.2 and 6 are met.

"ANNEX A" SENSORY AND PHYSICAL EXAMINATION

The sample used for sensory evaluation should not be same as that used for other examination.

1.   Examine the sample unit for foreign matter, bones and discolouration.

2.   Assess the odour in the uncooked sample in accordance with the Guidelines for the Sensory Evaluation of Fish and Shellfish In Laboratories (CAC/GL 31-1999).

3.   Assess the flavour in cooked sample in accordance with the Guidelines for the Sensory Evaluation of Fish and Shellfish In Laboratories (CAC/GL 31-1999).

4.   The sample shall be deep-fried in fresh cooking oil at 190oC for 20-60 seconds as appropriate to the thickness of the crackers.

SECTION 4

GUIDELINES FOR SENSORY EVALUATION OF
FISH AND SHELLFISH IN LABORATORIES

CODEX GUIDELINES FOR THE SENSORY EVALUATION OF FISH AND SHELLFISH IN LABORATORIES

CAC/GL 31 - 1999

GUIDELINES FOR THE SENSORY EVALUATION
OF FISH AND SHELLFISH IN LABORATORIES

I.   SCOPE AND PURPOSE OF THE GUIDELINES

The guidelines are intended to be used by analysts who need to apply sensory methods when using criteria based on sensory attributes of the products. Although the guidelines have been written with the Codex requirements in mind they include some provisions for products not covered by these standards but where sensory evaluation is used in the testing of fishery products for conformity requirements.[1] These guidelines are to be used for sensory examination of samples in a laboratory to determine defects by procedures, including cooking, which are not normally done by analysts in the field. Technical information is provided on the laboratory facilities used for such analyses and training of analysts.

The objective of guidelines is to ensure uniformity of application of standards by making recommendations for inspection purposes concerning the facilities required in sensory testing and the procedures for carrying out sensory tests.

For the purpose of this document the use of fish means finfish, crustaceans, and molluscs.

II.   FACILITIES FOR SENSORY EVALUATION

2.1   GENERAL OBSERVATIONS

Sensory evaluation should be carried out by adequately trained personnel (see Section IV). They evaluate a specialized range of products, and use one sensory methodology.

2.2   LABORATORIES FOR SENSORY EVALUATION

2.2.1   Location and Layout

Figure 1 illustrates a plan of a laboratory that would be suitable for use for examining fishery products. The plan illustrates the principle that the preparation area should be separate from the evaluation area.

Office accommodation, storage rooms, staff facilities, and possibly other test facilities should be provided elsewhere in the premises. The evaluation area must not be used for chemical and microbiological analyses however, some types of analyses could be done in the preparation area.

2.2.2   Preparation Area

This area is to used for the handling and storage of fishery products, and for the preparation of samples for sensory evaluation. It should be constructed so as to comply with the requirements of good manufacture practices for the design and construction of fishery establishments. The rooms should be designed to ensure cooking odours do not interfere with sensory analysis.

2.2.3   Evaluation Area

There should be no preparation of products in this area other than final trimming of samples prior to cooking.

The area, ventilation, procedures and sample sequence should be organized to minimize disturbing sensory stimuli. Also influence and disturbances from fellow evaluators and other personnel should be minimized. The colour of evaluation area should be neutral.

The working surfaces should be illuminated by daylight or artificial daylight. Any specific conditions in standards should be met.

---

[1] Additional criteria may be included if new recommendations are made by the Committee

Figure 1.   Illustrative Plan of a Laboratory for Sensory Evaluation of fishery Products

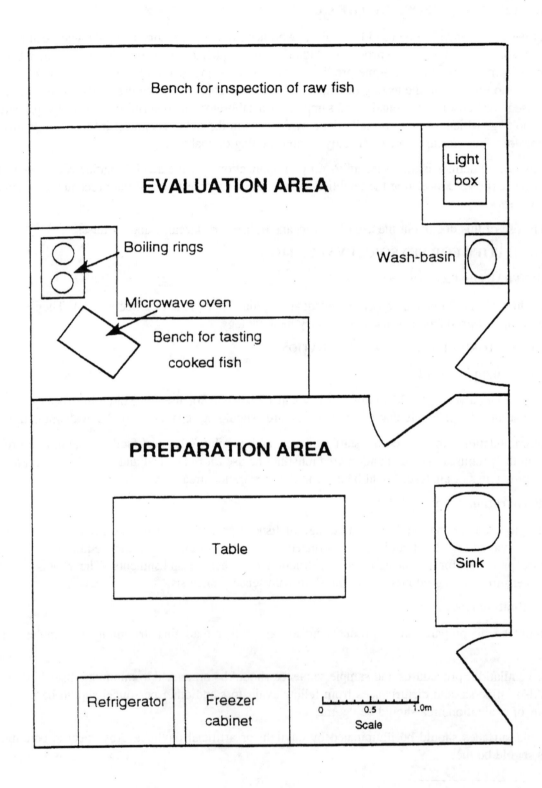

2.2.4 Equipment

The exact type and amount of equipment required will depend to some extent on the nature of products to be inspected and the volume and frequency of the examinations.

III. PROCEDURES FOR SENSORY EVALUATION

3.1 COLLECTING AND TRANSPORTING SAMPLES

In most circumstances where fishery products are subjected to sensory evaluation a decision is made about a batch of fish, for example, acceptance or rejection of a consignment of imported products, classification of batches of fish on a market into freshness grades. The decision is made on the basis of an examination of a sample drawn from the batch according to guidelines which will usually specify how a sample is to be taken for the intended regulatory or commercial purpose of the examination.

When collecting a sample for inspection the inspector should ensure that the procedures used for taking the sample, and the subsequent handling of the sample, do not materially affect its sensory properties.

The inspector should check that the sample is properly packed and where necessary, under temperature control before dispatching it to the inspection laboratory. If the sample is not under the supervision of officials during transport the inspector should ensure that sample can not be tampered with during the journey.

On receipt at the inspection laboratory, samples, if not evaluated immediately, should be stored under appropriate conditions. However fresh and chilled products should be examined on the day they are received. Products in either chill or frozen storage should be appropriately wrapped to prevent drying out or desiccation.

3.2 PREPARATION OF SAMPLES FOR EXAMINATION

Table 1 in Annex 1presents attributes useful in evaluating some species and products. Procedures for preparation of samples should be appropriate for the product types. Some procedures relative to fresh or frozen finfish are described in the following paragraphs.

The fish, if entire, should be gutted and the guts retained. The head should be removed, and the fillet from one side to taken off. The portions should be assembled on tray for analysis.

QF Products can be laid out on the examination bench in the evaluation area, but it is often more convenient for presentation and for clearing up after if sample units are presented on trays.

Frozen products should be first examined in the frozen state. The complete sample unit or portions of the unit should then be thawed for sensory evaluation. Whether the sample can, or should be subdivided, depends on the nature of the products. Packs of IQF shrimps or fillets can be opened and subsamples taken. Portions could be sawn off large fish or off blocks, but this might be difficult in the case of thick material unless a bandsaw is available.

Frozen material should be thawed out as quickly as possible, but without raising the temperature of all or part of the product so that it might spoil. The simplest procedure is to spread out the sample units on the benches and tables in the preparation area and leave them to thaw at ambient temperature. They should be covered to prevent drying and contamination. The progress of thawing should be monitored and when it is judged that thawing is complete the products should be evaluated, or transferred to a refrigerator. Products should be covered with plastic film before storing in the refrigerator. Storage should be limited in order to maintain sample integrity. If possible sample units should be thawed out on trays so that the amount and nature of the thaw drip can be assessed.

Thawing can be accelerated by immersion of the material in water. This is acceptable if product is protected from the contact with water by suitable wrappings, or if contact with water does not materially affect the sensory properties of the product. Care must be taken to prevent further spoilage or bacterial growth. Small sample units such as IQF fillets or small packs of shrimps or shellfish meats could be thawed in a microwave cooker on the defrost setting, but care must be taken not to use too high power settings otherwise parts of the material will be overheated.

Large frozen fish or large blocks of frozen products will take many hours to thaw out at ambient temperature, longer than a normal working day, and they can not be properly monitored throughout the whole process of thawing. One solution is to lay the products out for thawing at the end of a working day when they will just be completely, or almost completely, thawed by the following morning. Alternatively the material can be put out to thaw as early as possible in the day and transferred to a chill room at the end of the day to complete the process at low temperature. It is helpful to break apart blocks of product when they are partially thawed to accelerate thawing if this can be done without damaging the material.

3.3 COOKING

In cases where a final decision on odour or gelatinous state cannot be made in the thawed uncooked state, a small portion of the disputed material (approximately 200g) is sectioned from the sample unit and the odour and flavour or gelatinous condition confirmed by cooking without delay by one of the following cooking methods. The following procedures are based on heating the product to an internal temperature of 65 - 70°C. The product must not be overcooked. Cooking times vary according to the size of the product and the temperatures used. The exact times and conditions of cooking for the products should be determined by prior experimentation.

>*Baking Procedure*: Wrap the product in aluminum foil and place it evenly on a flat cookie sheet or shallow flat pan.

>*Steaming Procedure*: Wrap the product in aluminum foil and place it on a wire rack suspended over boiling water in a covered container.

>*Boil-in-Bag Procedure*: Place the product in a boilable film-type pouch and seal. Immerse the pouch in boiling water and cook.

>*Microwave Procedure*: Enclose the product in a container suitable for microwave cooking. If plastic bags are used, check to ensure that no odour is imparted from the plastic bags. Cook according to equipment instructions.[2]

3.4 PROCEDURES FOR THE ASSESSMENT OF PRODUCTS

Standards and specifications for fishery products will specify the features of the product that are able to be evaluated, and the criteria for accepting or rejecting products or for allocating them to grades. Table 1 presented in Annex I lists sensory attributes and criteria which may apply to standards and quality grading schemes. In order to apply quality criteria consistently in the inspection of products it is necessary to conduct the sensory assessments in a consistent and systematic manner. Samples should be assessed relative to the characteristics of the species concerned.

Assessors must pay particular attention to those features of the product which are referred to in any standards and which determinate conformance to the standard, but in addition they should assess and record other relevant attributes of the samples, as appropriate.

---

[2]   General Standard for Quick Frozen Fish Fillets, Annex A "Sensory and Physical Examination"

3.4.1   Assessment of Raw Products

Fresh fish will normally be assessed by appearance and odour. Fish change in appearance in a number of ways during spoilage in ice and it is not usually difficult to accurately grade iced fish by appearance alone. Characteristics to look for are listed in Table 1 of Annex I.

3.4.2   Assessment of Frozen Products

Frozen fish should be examined in the frozen state. The assessor should note the nature and state of any wrappings and glazes and the product should be examined for any discolourations and for the extent and depth of any dehydration. The assessor should note if there are any signs that product might have been thawed and refrozen. Signs of slumping or distortion of blocks, the collection of frozen drip in pockets in the wrappings, (not to be confused with water that might have been present on the fish at the time of freezing), and the partial loss of glaze.

Thawed samples should be presented and examined as for the corresponding unfrozen product where appropriate. It is not easy to evaluate the freshness of thawed whole fish by appearance because the freezing and thawing processes alter characteristics like the eyes, skin and colour of gills and blood. The gills have a leathery or slightly rancid odour even after short periods of frozen storage which have no significance for the quality of the product.

3.4.3   Assessment of Cooked Samples

Cooked samples should be held in a closed container, allowed to cool to a comfortable tasting temperature, and kept warm unless they are assessed immediately. Products which have already been cooked, for example cooked shrimps, should be warmed up slightly.

The assessor should note the appearance of the product and record any unusual features. The odour should be assessed and its character and strength recorded, particularly any unusual odours like chemical taints. Assessors should be encouraged to taste cooked samples as some compounds can only be detected by mouth (e.g. low levels of decomposition or fuel contamination).

The flavor of a sample in the mouth should confirm the assessment based on odour, but can give additional information. For example most additives such as salt, sorbates, polyphosphates, are not detectable by odour, but are detectable by taste. Sensory analysis alone should not be used to determine the presence of additives and any suspicion that non permitted additives have been used, or that excess amounts of permitted additives are present, should be confirmed by chemical analysis where appropriate.

IV.   TRAINING OF ASSESSORS

4.1   OBJECTIVE SENSORY TRAINING

4.1.1   Considerations for Objective Sensory Training

In the sections below examples are provided of test materials which have been used for screening and training analysts.

Objective sensory testing measures the intrinsic sensory attributes of a sample through the analytic sensory perceptions of human assessors. In order to conduct objective sensory analyses of fish and fish products, assessors must be selected for their ability to perform the sensory tasks required, must be trained in the application of the required test methods, and must be monitored for their ongoing ability to perform the sensory tasks. Thus, sensory training includes:

a)   The selection of assessors for basic sensory acuity and for the ability to describe perceptions analytically i.e. without the effect of personal bias. Allergies to seafood or to some food additives could eliminate an analyst candidate.

b)   The development of the analytical capability by familiarization with test procedures, improvement of ability to recognize and identify sensory attributes in complex food systems, and improvement of sensitivity and memory so that he/she can provide precise, consistent, and standardized sensory measurements which can be reproduced.

c)   The monitoring of the assessor's performance and the consistency of their analytic decisions by the frequent periodic assessment of the sensory decisions.

4.1.2   Selection of Candidate Assessors

A candidate for assessor training should demonstrate that he/she:

1.   is not anosmic (unable to perceive odours) - so that odours of decomposition and other defects will be perceived and described in a consistent manner;

2.   is not ageusic (unable to perceive basic tastes) - so that tastes associated with decomposition and other defects will be perceived and described in a consistent manner;

3.   has normal colour vision and is able to detect anomalies in the appearance of fish and fish products in a consistent manner,

4.   is able to rely on sensory perceptions and to report them appropriately;

5.   is able to learn terminology for new or unfamiliar perceptions (odours, tastes, appearances, textures) and to report them subsequently; and

6.   is able to define sensory stimuli and relate them to an underlying cause in the product.

The first five points can be measured in testing, the last ability is developed during specific product training.

In conducting the tests, it is useful to allow for repetition of the tests for basic taste and for odour perception. This is necessary to ensure that the candidate is being tested for basic ability and not responding to an unfamiliar testing situation. New code numbers and presentation sequences are used in each test method.

4.1.2.1 Screening for Perception of Basic Tastes

The diversity of flavours, especially of defects from decomposition, which the inspector will be required to perceive and describe make it essential that some indication of the general ability to perceive basic tastes be established. One area of particular importance in selection and training is the ability to discriminate bitter and sour tastes/flavours as this is a common area of confusion in inexperienced assessors. These tastes/flavours are critical in the examination of fish and fish products as they are evident in the early stages of decomposition.

A matching standards test using concentrations which should be perceived by a normal taster has been described by several standard sources. The concentrations used have been shown in testing to be perceptible.

TABLE 1   **A SELECTION OF PUBLISHED TEST SOLUTIONS USED FOR SCREENING AND TRAINING ANALYSTS**

Basic Tastes	Standard Compounds Used (in water)	DFO Screening Tests (1986-96)	Meilgaard et al. (slight to very strong) (1991)	Jellinek (1985)	ASTM (1981)	Vaisey Genser and Moskowitz (1977)
Bitter	caffeine	0.06%	0.05 to 0.2%	0.02 & 0.03%	0.035, 0.07 & 0.14%	0.150%
Sour	citric acid	0.06%	0.05 to 0.20%	0.02, 0.03 & 0.04%	0.035, 0.07 & 0.14%	0.01%
Salt	sodium chloride	0.02%	0.2 to 0.7%	0.08 & 0.15%	0.1, 0.2% 0.4%	0.1%
Sweet	sucrose	2.0%		0.40 & 0.60%	1.0, 2.0 & 4.0%	1.0%
umami*	monosodium glutamate	0.08%				

* This has been identified by some analysts as being a fifth basic taste, however this remains controversial. This **may** be used as part of the selection procedure, but should definitely be used as part of the training sessions to illustrate the contribution to the flavours of fish contributed by the ribonucleotides.

4.1.2.2 Screening for Perception of Odours

In this case, several types of tests are available which will accomplish the selection procedure.

Because people are able to perceive a very large number of separate odour qualities, the samples used should be chosen to be both representative of common odours with which the candidate would likely have had experience, and also be representative of odour qualities which occur as defects in fish and fish products. Two examples two test methods which would be appropriate for use in assessing odour perception are presented in Annex II.

4.1.2.3 Screening of Normal Colour Perception

Colour blindness is measured by the use of one of several standard ophthalmologic tests including the Ishihara Colour Blindness Test and the Farnsworth-Munsell 100-Hue Test. These tests may be purchased through medical supply sources and should come with complete instructions as to their use. They must be administered under the exact conditions specified in the instructions.

4.1.2.4 Screening Test for the Assessment of Texture

There can be cases when fish is rejected for texture. These are tests which are essentially done by touch on raw product. Characteristics which may be assessed include:

    a)   firmness: in fresh fish and shellfish (shrimp); and

    b)   springiness: in fresh fish.

One such test is the procedure designed by Tilgner (1977) and reported in Jellinek (1985). This test used a series of samples which increase slightly in firmness and uses pressure with the forefinger of the dominant hand to assess firmness and allow the candidate to rank the samples from least to most firm. This allows the assessment of the concept of firmness and the concept of increasing intensity in a sensory attribute. The samples used in the test described are permanent samples cast from polyvinyl chloride although a series of samples can also be generated from appropriate food samples.

4.1.3 Training of Assessors

A *Suggested Syllabus for a Training Course for Assessors in the Sensory Assessment of Fish and Fishery Products* The following is a model training syllabus. The length of the basic sensory science training which is included in the course can vary from the 10 hours (1.5 days) shown below to full length courses of university level training. It is suggested that hands-on exercises accompany each section to demonstrate the concept under discussion (e.g. prepare basic taste solutions and have the students taste them during the lecture on taste). A Suggested Syllabus for a Training Course for Assessors in the Sensory Assessment of Fish and Fishery Products is presented in Annex III.

4.1.4 Monitoring of Assessors

The validation of the effectiveness of sensory training and of the consistency of sensory assessments is achieved through ongoing monitoring of the sensory decisions made by the assessor. This may be accomplished in a variety of ways, either singly or in combination.

a)   The first is the use of check samples which are samples of known quality which are distributed to inspectors for examination in their day-to-day testing facility. The results are sent back to the central coordinator of the samples for analysis. The advantage of this method is that samples are being assessed under the actual laboratory conditions. Samples used for this are prepared using the procedures described in Section 4.2, Preparation and Handling of Samples. Also commercial product of known quality and which is available in sufficient quantity may be used.

b)   Another procedure which is used to validate the performance of an inspector is through actual accreditation testing and calibration procedures. These are conducted in a central location laboratory which is large enough to accommodate all of the inspectors participating in the test. Samples are prepared using the procedures described in Section 4.2 Preparation and Handling of Samples. Also commercial product of known quality and which is available in sufficient quantity may be used. This procedure must be repeated at regular intervals to ensure that no change has occurred in the inspectors' ability to evaluate products and the inspector must reach a pre-defined level of performance on both «pass/accept» samples and «fail/reject» samples.

c)   A supplementary method of evaluation of an inspector's performance is the accumulation over time of the on-going inspection results vs. any other known information on samples, e.g. reinspection results, consumer complaints, chemical analyses, etc.

4.1.5 Reference Documents

Reference documents are presented in Appendix II.

4.2 PREPARATION AND HANDLING OF SAMPLES

4.2.1 Type of Samples

Samples used for the purpose of training individuals in sensory techniques concerning fishery products are the single most important factor to be considered. It is imperative that proper samples be provided in reference to sensory training.

There are two types of samples to be considered in the training of sensory analysts or inspectors.

1.   Controlled spoilage samples: These samples should display or represent a full range of quality, as well as the normal range of product characteristics related to odour, flavor, appearance, and texture.

It is essential that samples of excellent quality be provided as a reference point during the preparation of such packs.

Quality defects should be naturally occurring, if possible, to exhibit sensory characteristics which are typical of the product to be used. If the samples are spoiled or contaminated artificially, they may not exhibit typical sensory properties for both the acceptable or unacceptable units to be used for training.

It is important for the individual preparing the samples to have knowledge of the normal commercial processing of the product to be spoiled from harvesting to freezing and be aware of processing methods and conditions under which spoilage usually occurs. Understanding the general pathways of decomposition would be useful in the preparation of controlled spoilage samples.

When possible, controlled spoilage, samples should be prepared where the product is harvested and processed to allow for the species, flora, etc. to duplicate normal spoilage conditions that allows for typical odours of decomposition as well as other characteristics that mimic commercial samples.

2.   Commercial samples: Whenever possible, the use of commercial samples should be incorporated into the sensory training of individuals. Many times, quality defects (odour, flavor, appearance, texture, etc.), as well as taints (musty/mouldy odours, flavours, rancidity, petroleum distillates, etc.) can be best shown with commercially produced samples that have these defects. These commercially manufactured samples allow one to assess sensory personnel during training by providing «real life» samples. They can also be used to measure an individual's retention abilities as it relates to making correct decisions in sensory science.

Many times, quality defects and taints are not found in all intensities in controlled spoilage samples but can be shown in slight, medium, and strong intensity from commercially produced samples.

4.2.2 Preparation of Sample Packs

Sample preparation should be started in plenty of time to allow one to obtain the majority of defects as well as allowing product to go through a curing process if necessary.

If possible, the spoilage run should be conducted with fish «in the round» to allow for natural spoilage to occur. This allows for typical spoilage odours to form.

(1)   Baseline

It is essential that excellent quality material of all species and product forms of known history, without commercial abuse, be obtained to provide a constant reference to the workshop participants. Whenever possible, both fresh and frozen product forms should be included in the preparation of controlled spoilage samples. The lot should be uniform with respect to its quality at the start of the run.

Proper record keeping is essential in the preparation of spoilage samples. Samples of each code taken should be consistent within a set, each succeeding set representing a longer period of time that the product has been held under ambient or iced conditions. Temperature monitoring is essential to prevent fluctuations during each spoilage run.

Spoilage must be accomplished under appropriate conditions of temperature and environmental contamination if authentic spoilage effects are to be obtained. Variations in spoilage rates between individual units can be minimized if the starting material is of uniform size and quality and contact between individual units is maintained during spoilage.

Fish tend to spoil at different rates so one should examine product at regular intervals and group the product together that have similar characteristics prior to processing. Expert evaluation of the samples is constantly needed at this stage.

The number of increments needed will depend on the purpose of the training and the species to be examined but a minimum of 5 increments and as many as 8 may be needed. At least 50% of the pack should be of acceptable product.

(2)   Spoilage

Generally, both high and low temperature decomposition spoilage should be included, but knowledge of the species and the standard processing method and at what point of the process is spoilage most likely to occur should determine the general spoilage method. It is important to avoid «shortcuts» for the sake of convenience. If pre-chilling spoilage is the issue, the use of frozen fish must be avoided. Careful temperature control is a necessity.

(3)   Packaging and Storage

The species and type of product from a spoilage run should be taken into account to determine the amount of shelf life one can expect.

Canned products should be allowed to cure in the can for at least 30 days prior to use. They should be stored in a cool and dry location with a temperature range of 14°C - 18°C, otherwise one can expect a much shorter storage life. Maximum shelf life of canned seafood products for training purposes is approximately 2 years. After this amount of time, characteristics develop that may affect one's judgement or render the samples of little value for training purposes.

Unless freezer storage damage is intended to be demonstrated, raw and pre-cooked frozen products should be properly glazed to prevent dehydration and freezer burn. Depending on the length of storage, the samples may require periodic reglazing to ensure the quality. If possible, product should be vacuum packed to ensure quality and is essential in the storage of some fish species as well as pre-cooked samples.

Both raw, precooked and canned controlled spoilage samples should be evaluated by a qualified individual prior to use in a workshop. The samples should have both chemical analysis and sensory results to determine the quality of the increment and the homogeneity of the increment.

4.2.3   Characteristics of Samples

4.2.3.1 Sensory Attributes

A.   Must show normal odour, flavor, appearance, texture, etc. characteristics of the species to be used for samples.

B.   If product forms normally show characteristics attributed to harvest location, feed odours, etc. include with the controlled spoilage samples if possible.

C.   Samples which exhibit odours of spoilage or contamination defects must not be too intense to the point of overpowering the participant's senses and affecting judgement of other samples during a training session.

D.   Samples showing slight to moderate odours of spoilage or contamination provide more of a challenge and better represent «real world» conditions.

E.   Each increment or code must show consistent or similar characteristics to have value when used for training.

4.2.3.2 Chemical Attributes

Inclusion of chemical attributes of authentic pack samples can be useful in training (see Annex III Section II Practical Exercises from the model Syllabus).

A.   Chemical indicators of decomposition (CID) are selected that are essentially absent in the fresh product.

B.   A CID is selected that will monitor the decomposition pathway of interest in the particular products to be used for training. Methods are used which are capable of differentiating between the CID levels found in passable, slightly abused-passable and the first definite stage of decomposition. When possible it is preferable to use two CID's.

C.   The CID should be retained in the processed forms (washed/cooked/canned/stored) of the fishery product to be examined.

D.   The changes in a CID should track the changes in sensory quality in the fishery product.

E.   A sufficient number of subsamples should be analyzed for each increment of prepared sample to measure the degree of variation within sample increments. This is especially important for those increments representing the transition from a passable product to the first definite stage of decomposition.

TABLE 1.   *EXAMPLES OF ATTRIBUTES OF FISHERY PRODUCTS USED IN SENSORY EVALUATION*[3]

Presentation	Feature	Criteria and description
	Vertebrate fish, iced	
Raw whole, gutted or ungutted	outer surface, skin	colour: bright, dull, bleached slime: colourless, discoloured damage: none, punctures, abrasions
	eyes	shape: convex, flat, concave brightness: clear, cloudy colour: normal, discoloured
	belly cavity	guts (in intact fish): intact, digested cleanliness (in gutted fish): completely gutted and cleaned, incompletely gutted, not washed belly walls: bright, clean, discoloured, digested parasites: absent, present blood: bright, red, brown
	texture, appearance of gills	skin: smooth, gritty, flesh, firm, soft colour: bright red or pink, beached, discoloured mucus: clear, opaque, discoloured
	odour of gills	fresh, characteristic, neutral, slightly sour, slightly stale, definite spoilage, putrid
Raw fillets	appearance	translucent, glossy, natural colour, opaque, dull, blood-stained, discoloured
	texture	firm, elastic, soft, plastic
	odour	marine, fresh, neutral, sour, stale, spoiled, putrid
Cooked fillets	odour	spoilage: marine, fresh, neutral, musty, sour, spoiled taints: absent, disinfectant, fuel oil, chemicals, sulphides
	flavour	spoilage: sweet, creamy, fresh oil, neutral, sour, oxidised, putrid, musty, fermented, rancid, bitter, taints: absent, disinfectant, fuel oil, very bitter, alkaline, polyphosphates, chemicals
	texture	succulent, firm, soft, pasty, gelatinous, dry

---

[3]   References to be included for the clarification of sensory properties, as established by ISO

Vertebrate fish, frozen

Frozen	appearance	freezer burn: absent, slight, superficial, extensive, deep colour: normal, yellow to bronze discolouration in fatty fish
Thawed fillets, raw	texture	firm, elastic, flexible, very firm, hard, stiff drip: slight, moderate, abundant odour spoilage and taints: as for chilled fish cold storage: absence of cold storage odours, sharp, cardboardy, rancid
Thawed fillets	Odour and flavour	spoilage and taints: as per chilled fish cold storage: absence of cold storage odours, sharp, cardboardy, rancid
	texture	firm, succulent, tough, fibrous, dry

Crustacean shellfish, chilled

Raw	appearance, shell on	bright colours, slight blackening on the head, blackening on head and body
	appearance, peeled meats	translucent, overall white or light grey, slight black discolouration, extensive black discolouration, very translucent, slimy, yellowish discolouration on butt end of tail meat taken from head-on products
	odour	fresh, marine, musty, ammoniacal, sour, spoiled, putrid
Cooked meats	appearance	white, opaque, blacks spots, extensive back discolouration, slightly translucent
	odour	fresh, boiled milk, musty, ammoniacal, rancid, sour, spoiled
	flavour	sweet, creamy, neutral, musty, sour, bitter, spoiled
	texture	firm, elastic, soft, mushy

Crustacean shellfish, frozen

Criteria specific to the grading of frozen shellfish, and their descriptions, are essentially the same as those applied to the grading of frozen vertebrate fish.

Cephalopods, fresh or refrigerated

colour	skin: bright, dull, bleached meat: pearly white, lime coloured, pinkish or light yellow
adherence	adherent to the meat, easily separating from the meat
texture	meat: very firm, firm, slightly soft tentacles: resistant to tearing off, can be torn off easily
odour	fresh, seaweed, slight or no odour, sour

ANNEX II

EXAMPLES OF TEST METHODS WHICH WOULD BE APPROPRIATE FOR USE IN SCREENING ASSESSORS FOR ODOUR PERCEPTION

1.   The following is a list of samples as used in Canada:

(a)   canned salmon (fish)

(b)   canned sardines (fish/smoke)

(c)   yeast (growth of yeasts)

(d)   coffee (common product - to illustrate the method)

(e)   orange & pineapple (fruity odours)

(f)   cucumber & asparagus (vegetable odours)

(g)   vinegar, cinnamon, pepper & cloves (pungent odours which can be differentiated)

(h)   vanilla (sweet odour)

(i)   prepared mustard (strong vinegar component, illustrates ability to perceive in mixtures)

(j)   acetone, rubbing alcohol (contaminants, solvents)

(k)   petroleum product (fuel oils)

(l)   old vegetable oil (rancid oil)

In this test, the candidate is asked to identify the samples only by the odour as all visual information is masked. The sample are then identified and discussed with the candidate and the number of correct identifications recorded. During this step the candidate is given the opportunity re-examine any of the samples. The test is repeated after a time period such as 2 or 4 hours (during which other selection tests or interviews may be given), and number of correct responses recorded. The improvement in test scores which should occur (unless all were correct on the first round) gives an indication of the ability of the candidate to learn new terms to describe sensory perceptions.

2.   The University of Pennsylvania Smell Identification Test, a standardized test for assessment of odour perception, is available from Sensonics, Incorporated, 155 Haddon Avenue, Haddonfield, New Jersey, 08033 USA.

SUGGESTED SYLLABUS FOR A TRAINING COURSE FOR ASSESSORS IN THE SENSORY ASSESSMENT OF FISH AND FISH PRODUCTS

I.   LECTURES

Part I: Theoretical Principals and Laboratory Practices of Sensory Assessment (10 Hours)

A.   Basic Sensory Testing Principles:

    1.   Affective or subjective testing (test types, information gained, data collection, respondent type and numbers, decision-making possible from this information).

    2.   Analytical or objective testing (test types, information gained, data collection, respondent type and numbers, decision-making possible from this information).

        i.   Discriminative testing: types of information that is gained and that is not.

        ii.   Descriptive testing: qualitative and quantitative.

    3.   The role of the fish and seafood assessor or product expert in sensory testing.

B.   Action of the Senses and the Perception of Sensory Properties of Fishery Products:

    1.   The physiology of the senses - sight, smell, taste, touch and hearing;

    2.   The perception of sensory properties - appearance/colour, odour, flavor, texture; and

    3.   Sensory interactions.

C.   Sample Evaluation techniques:

    1.   Odour evaluation techniques.

    2.   Flavor evaluation techniques.

    3.   Texture evaluation (firmness and springiness).

    4.   Special techniques for seafood samples.

D.   Basic Psychophysics of Sensory Assessment:

    1.   Thresholds; detection and recognition.

    2.   Intensity; the logarithmic nature of character strength perception.

    3.   Saturation; explanation of the phenomenon.

E.   Factors Influencing Sensory Judgements:

    1.   Physiological effects; blending; masking, carry-over, enhancement and suppression.

    2.   Psychological effects; expectation, stimulus, halo, order, proximity, stimulus, logical, suggestion, contrast and convergence, and central tendency.

    3.   Control of physiological and psychological effects.

F.   Basic Data Collection and Analysis:

    1.   Discriminative methods: triangle (3-alternative forced choice or balanced design), duo-trio, two-out-of-five, paired comparison):

        i.   Ballot information and design types

        ii.   Analysis of data

    2.   Descriptive methods: Flavor Profile, Texture Profile, Spectrum, QDA:

        i.   Scales; category, line, magnitude estimation

        ii.   Ballot information and design types

        iii.   Analysis of data

    3.   Sensory methods for quality control - general discussion.

G.   Terminology and the use of reference standards. the analyst should «understand the role of sensory descriptors as an aid to developing long term sensory memory and as a means of communicating results». (see Appendix 1):

    1.   Terminology development (including internationally recognised sources for known terms).

    2.   The importance of definitions

    3.   The use of reference Standards

    4.   Overview of terms relevant to seafood quality, with specific attention to those associated with low levels of decomposition.

H.   Sample Handling and Preparation:

    1.   Presentation and coding.

    2.   Randomization of samples; purpose and occasion for use.

    3.   Homogeneity of samples and serving temperature.

    4.   Sample size and quantity.

Part II: **Deterioration of Fish and Fish Products (3 Hours)**

A.   Composition of Fish and Shellfish:

    1.   Major components: protein, fat, carbohydrate, water.

    2.   Minor components; non-protein nitrogenous compounds, minerals, vitamins.

B.   Pathways of Quality Deterioration:

    1.   Breakdown of protein, fat, non-protein nitrogenous compounds, and, for some species, carbohydrates.

    2.   Microbial spoilage.

    3.   Terminology associated with each type of spoilage pathway.

C.   Chemical Indicators of Fish Quality and the Correlation of these with Sensory Data.

Part III: **Contamination and Taint (1 Hour)**

A.   Types:

1.   Naturally-occurring (muddy-earthy off-flavours).

2.   Man-made (petroleum, pulp and paper effluent, other processing effluents).

B.   Mechanism of flavor and odour changes.

C.   Testing methods for contamination and/or taint (special considerations).

II.   PRACTICAL EXERCISES

Part I: Presentation of Seafood Related Terminology, Clear Definitions, and References Which Demonstrate the Terms (2 hours)

Part II: Spoilage and Decomposition (18 hours)

This portion of the course provides hands-on experience. It is suggested that only one species at a time be evaluated.

This section may include whole fish, fillets, canned fish and/or smoked fish and other specialty products. Whenever possible, trainees should evaluate flavor as well as odour, e.g. especially in products such as canned fish packed in oil as the packing medium can mask odours.

The following sequence of three session formats are suggested for each species and will require approximately 4 hours in total. It is suggested that the effectiveness of the training be evaluated by testing the trainee's ability to assess sample quality correctly before moving on to another species:

(a)   Demonstration session: Group demonstrations of samples of known quality by an experienced product expert. The labelled samples should represent a full range of quality, in order from highest to lowest quality, with discussion of sensory results, descriptors, as well as any data from chemical indicators of quality which are appropriate for that species.

(b)   Discussion session: Random presentation of blind-coded samples for individual evaluation an group discussion of the results.

(c)   Testing session: individual evaluation of blind-coded test samples and comparison of results with product expert.

The collection and analysis of data with detailed discussions of the samples will provide feed-back to the trainees.

Part III: Deterioration in Frozen Stored Fish and Shellfish (4 hours)

A.   Demonstration of varying degrees of defects in appearance, odour, flavor, and texture caused by frozen storage of seafood products.

B.   Include both low-fat and high-fat fish and seafood samples.

C.   Have available terminology, definitions, and references for the oxidation process and for textural changes.

Part IV: Deterioration in Canned Fish and Shellfish (4 hours)

A.   As for section II, and also to include information on pre- and post- processing deterioration.

Part V: Other Defects (2 hours)

A.   Detection of taints using spiked samples (assess by odour only).

B.   Demonstration of visual defects.

DEFINITIONS OF SOME OF THE TERMS USED IN SENSORY
ANALYSIS OF SEAFOOD

Appearance	All the visible characteristics of a substance/sample
Analyst/ Assessor	Any person taking part in a sensory test
Bilgy	The aromatic associated with anaerobic bacterial growth, which is illustrated by the rank odour of bilge water. The term «bilgy» can be used to describe fish of any quality which has been contaminated by bilge water on board a vessel. Bilge water is usually a combination of salt water fuel, and waste water
Bitter	One of the four basic tastes, primarily perceived at the back of the tongue, common to caffeine and quinine. There is generally a delay in perception (2-4 seconds)
Briny	The aroma associated with the smell of clean seaweed and ocean air
Chalky	In reference to texture, a product which is composed of small particles which imparts a drying sensation in the mouth. In reference to appearance, a product which has a dry, opaque, chalk like appearance
Cucumber	The aroma associated fresh cucumber, similar aromas can be associated with certain species of very fresh raw fish
Decompose	To break down into component parts
Decomposed	Fish that has an offensive or objectionable odour, flavor, colour, texture, or substance associated with spoilage
Distinct	Capable of being readily perceived
Feedy	«Feedy» is used to describe the condition of fish that have been feeding heavily. After death, the gastric enzymes first attack the internal organs, then the belly wall, then the muscle tissue. If the enzymes have penetrated into the flesh, they are capable of causing quality changes dimethyl (DMS), and may be attributed to certain zooplankton as it passes through the food chain. The odour of «feed» fish has been described as similar to certain sulfur containing cooked vegetables, such as broccoli, cauliflower, turnip, or cabbage
Fecal	Aroma associated with feces
Firm	A substance which exhibits moderate resistance when force is applied in the mouth or by touch;
Fish	Means any of the cold-blooded aquatic vertebrate animals commonly known as such. This includes Pisces, Elasmobranchs and Cyclostomes. Aquatic mammals, invertebrate animals and amphibians are not include
Fishy	Aroma associated with aged fish, as demonstrated by trimethylamine (TMA)or cod liver oil. May or may not indicate decomposition, depending on species
Flavor	An attribute of foods resulting from the stimulation of taste, smell, sight, pressure, and often warmth, cold or mild pain
Freshness	Concept relating to time, process, or characteristics of seafood as defined by a buyer, processor, user, or regulatory agency

Fruity	Aroma associated with slightly fermented fruit. Term is used to describe odours resulting from high temperature decomposition. Example = canned pineapple
Gamey	The aroma and/or flavor associated with the heavy, gamey characteristics of some species such as mackerel. Similar to the relationship of fresh duck meat as compared to fresh chicken meat
Glossy	A shiny appearance resulting from the tendency of a surface to reflect light at 45 degree angle
Grainy	A product in which the assessor is able to perceive moderately hard, distinct particles. Sometimes found in canned seafood products
Intensity	The perceived magnitude of a sensation
Iridescent	An array of rainbow like colours, similar to an opal or an oil sheen on water
Intensity	The perceived magnitude of a sensation
Masking	The phenomenon where one sensation obscures one or several other sensations present
Mealy	Describes a product that imparts a starch-like sensation in the mouth
Metallic	Aroma and/or taste associated with ferrous sulphate or tin cans
Moist	The perception of moisture being released from a product. The perception can be from water or oil
Mouldy	Aroma associated with mouldy cheese or bread
Mouth coating	The perception of a film in the mouth
Mouth filling	The sensation of a fullness dispersing throughout the mouth. A umami sensation, as stimulated by MSG
Mushy	Soft, thick, pulpy consistency. In seafood little or no muscle structure discernible when force is applied by touch or by mouth
Musty	The aroma associated with a mouldy, dank cellar. Product can also have a musty flavor
Odour	Sensation due to stimulation of the olfactory receptors in the nasal cavity by volatile material. Same as aroma
Off odour	A typical characteristics often associated with deterioration or transformation of a flavor product
Opaque	Describes product which does not allow the passage of light. In raw muscle tissue of fishery products, this is usually due to the proteins loosing their light reflecting properties due to falling pH
Pasty	A product which sticks together like paste in the mouth when mixed with saliva. Forms a cohesive mass which may adhere to the soft tissue surfaces of the mouth or fingers
Persistent	Existing without significant change; not fleeting
Pungent	An irritating, sharp, or piercing sensation
Putrid	Aroma associated with decayed meat
Quality	A degree of excellence. A collection of characteristics of a product that confers its ability to satisfy stated or implied needs
Rancid	Odour or flavor associated with rancid oil. Gives a mouth-coating sensation and/or a tingling perceived on the back of the tongue. Sometimes described as «sharp» or «painty»
Reference	Either a sample designated as the one to which others are compared, or another type of material used to illustrate a characteristic or attribute
Rotting vegetable	Aroma associated with decayed vegetables, in particular the sulfur containing vegetables, such as cooked broccoli, cabbage, or cauliflower

Rubbery	A resilient material which may be deformed under pressure, but returns to its original form once the pressure is released
Salty	The taste on the tongue associated with salt or sodium
Sensory	Relating to the use of the sense organs
Slimy	A fluid substance which is viscous, slick, elastic, gummy, or jelly-like
Sour	An odour and/or taste sensation, generally due to the presence of organic acids
Stale	Odour associated with wet cardboard or frozen storage. Product can have a stale flavor as well
STP	Sodium tripolyphosphate. Can produce a soapy, alkaline feel and taste in the mouth
Sweet	The taste on the tongue associated with sugar
Taste	One of the senses, the receptors for which are located in the mouth and activated by compounds in solution. Taste is limited to sweet, salty, sour, bitter and sometimes umami
Terminology	Terms used to describe the sensory attributes of a product
Translucent	Describes an object which allows some light to pass, but through which clear images can not be distinguished
Transparent	Describes a clear object, which allows light to pass and through which distinct images appear
Umami	Taste produced by substances such as monosodium glutamate (MSG) in solution. A meaty, savory, or mouth filling sensation
Watermelon	Aroma characteristic of fresh cut watermelon rind. Similar odours are sometimes found in certain species of very fresh raw fish
Yeasty fermented	Aroma associated with yeast and fermented products such as bread or beer

APPENDIX 2

REFERENCE DOCUMENTS

ASTM Atlas of odour character profiles, publication DS 61, PCN 05-061000-36. Complied by Andrew Dravnieks.

ASTM Committee E-18, 235, draft of terminology document.

ASTM Aroma and Flavor Lexicon for Sensory Evaluation DS 66. G.V. Civille and B.G. Lyon, eds.

ASTM Committee E-18 on Sensory Evaluation of Materials and Products, 1981. STP 758 - Guidelines for the Selection and Training of Sensory Panel Members.

ASTM Committee E-18 on Sensory Evaluation of Materials and Products, Terminology Committee, (date?). Draft definition for «Expert» and «Expert Assessor».

Cardello, A. 1993. Sensory methodology for the classification of fish according to edibility characteristics. *Lebensmittel-Wissenschaft-und-Technologie* 16, 190-194.

Department of Fisheries and Oceans, Canada. Code of practice for fishery products.

Department of Fisheries and Oceans, Canada. Regulations respecting the inspection of processed fish and processing establishments.

Department of Fisheries and Oceans, Canada, Inspection Branch. 1986 to 1995. Notes from «*Sensory Methods in Fish Inspection*» - Sensory Training course given by the National Centre for Sensory Science, Inspection Branch, Department of Fisheries and Oceans, Canada.

Howgate, Peter 1992. Codex review on inspection procedures for the sensoric evaluation of fish and shellfish. CX/FFP 92/14.

IFST - International Institute of Food Science and Technology. «Sensory Quality Control: Practical Approaches in Food and Drink Production». Proceedings of a joint symposium at the U. of Aston, 6-7-January, 1977. Session II, «Measurement of Fish Freshness by an Objective Sensory Method». P. Howgate, p. 41.

ISO 5492 (1983) Sensory analysis - vocabulary.

ISO 8586-2 Sensory Analysis - General guidance for the selection, training and monitoring of assessors - Part 2. Experts

Jellined, G. 1985. *Sensory Evaluation of Food - Theory and Practice*. Ellis Horwood, Ltd., Chichester, England.

Johnsen, et al., 1987. A lexicon of pond-raised catfish flavor descriptors. J. Sensory Studies 4, 189-199.

Laverty, 1991. «Torry Taste Panels». In Nutrition and Food Science, Vol 129 No. 2-4. Includes terminology based on odour of gills in raw, iced cod.

Learson, Robert 1994, personal correspondence. NOAA/NMFS Research Laboratory, Gloucester, MA.

Multilingual guide to EC freshness grades for fishery products. Torry research station, Aberdeen, Scotland and the West European Fish Technologists Association (WEFTA). Compiled and edited by P. Howgate, A. Johnston, and K.J. White.

NOAA Handbook 25, part 1, Inspection.

NOAA/NMFS, Technical Services Unit.

Kramer and Liston, (eds) Seafood Quality Determination. Proceedings of the International Symposium on Seafood Quality Determination, Coordinated by the University of Alaska Sea Grant College Program, Anchorage. Alaska, 10-14 November, 1986.

Learson and Ronsivalli, (1969), A new approach for evaluating the quality of fishery products.

Meilgaard, M., Civille, G.V., and Carr, B.T. 1991. Sensory Evaluation Techniques. CRC Press, Inc., Boca Raton, FL.

Poste, L., Mackie, D., Butler, G. and. Larmond, E. 1991. Laboratory Methods for Sensory Analysis of Food. Agriculture Canada Research Branch.

Prell and Sawyer, 1988 «Flavor Profiles of 17 Species of North Atlantic Fish» J. Food Science, 53, 1036-1042.

Prell and Sawyer (1988). Consumer evaluation of the Sensory Properties of Fish» J. of Food Science 53, 12-28, 24.

Reilly, T.I. and York, R.K. 1993. Sensory analysis application to harmonize expert assessors of fish products. Proceedings of «Quality Control and Quality Assurance of Seafood», May 16-18, 1993, Newport, Oregon (Eds. Sylvia, G., Shriver, A.L. and Morrisey, M.T.)

Sawyer et al.., (1988) «Consumer evaluation of the sensory properties of fish». J. of Food Science, Vol. 53. No. 1

Sawyer, F.M. et al. 1981. A comparison of flavor and texture characteristics of selected underutilized species of North Atlantic fish and certain treatment of fish. International Institute of Refrigeration. Paris, France. p. 505.

Shewan et al., (1953), The development of a numerical scoring system for the sensory assessment of the spoilage of wet white fish stored in ice. J. Sci. Food Agric., 4 June.

Soldberg, et al. (1986), Sensory profiling of cooked, peeled and individually frozen shrimp». In Seafood Quality Determination, Elsevier Science Publishers.

Vaisey Genser, M. and Moskowitz, H. R. 1977. Sensory Response to Food. Forster Publishing Ltd., Zurich, Switzerland.

Wilhelm, Kurt, 1994, personal correspondence. NOAA/NMFS Research Laboratory, Gloucester, MA.